Towards Green ICT

RIVER PUBLISHERS SERIES IN COMMUNICATIONS

Volume 9

Consulting Series Editors

MARINA RUGGIERI
University of Roma "Tor Vergata"
Italy

HOMAYOUN NIKOOKAR
Delft University of Technology
The Netherlands

This series focuses on communications science and technology. This includes the theory and use of systems involving all terminals, computers, and information processors; wired and wireless networks; and network layouts, procontentsols, architectures, and implementations.

Furthermore, developments toward new market demands in systems, products, and technologies such as personal communications services, multimedia systems, enterprise networks, and optical communications systems.

- Wireless Communications
- Networks
- Security
- Antennas & Propagation
- Microwaves
- Software Defined Radio

For a list of other books in this series, see final page.

Towards Green ICT

Editors

Ramjee Prasad

CTIF, Aalborg University, Denmark

Shingo Ohmori

CTIF-Japan, Yokosuka Research Park, Japan

Dina Šimunić

University of Zagreb, Croatia

LONDON AND NEW YORK

Published 2010 by River Publishers
River Publishers
Alsbjergvej 10, 9260 Gistrup, Denmark
www.riverpublishers.com

Distributed exclusively by Routledge
4 Park Square, Milton Park, Abingdon, Oxon OX14 4RN
605 Third Avenue, New York, NY 10158

First published in paperback 2024

Towards Green ICT / by Ramjee Prasad, Shingo Ohmori, Dina Šimunić.

Routledge is an imprint of the Taylor & Francis Group, an informa business

Publisher's Note
The publisher has gone to great lengths to ensure the quality of this reprint but points out that some imperfections in the original copies may be apparent.

While every effort is made to provide dependable information, the publisher, authors, and editors cannot be held responsible for any errors or omissions.

ISBN: 978-87-92329-34-9 (hbk)
ISBN: 978-87-7004-550-6 (pbk)
ISBN: 978-1-003-33986-1 (ebk)

DOI: 10.1201/9781003339861

To the organisers of the 12th International
Symposium on Wireless Personal Multimedia
Communications (WPMC'09)

Table of Contents

Preface   xi

Acknowledgements   xv

List of Abbreviations   xvii

Part I: Overview

1. Contribution to Green Communications Vision   1
   D. Šimunić, R. Prasad and S. Ohmori

2. How Telecom Could Save the Planet   29
   L.E. Mägi, M.G. Gustafsson and A. Kramers

3. Directions towards Future Green Internet   37
   H. Imaizumi and H. Morikawa

4. Trend and Technology Development in Green Communications: KDDI's Technical Approach   55
   T. Mizuike and K. Sugiyama

5. Role of ICT in Positively Impacting the Environment   65
   K. Ramareddy, P. Pruthi and R. Prasad

6. Green Mobile   75
   K.E. Skouby and I. Windekilde

7. ZigBee as a Key Technology for Green Communications   87
   C. Spiegel, S. Rickers, P. Jung, W. Shim, R. Lee and J.H. Yu

8. Using ICT in Greening: The Role of RFID   97
   R. Krigslund, P. Popovski, I. Dukovska-Popovska,
   G.F. Pedersen and B. Manev

Part II: Energy/Power Optimization

9. Energy-Efficient Deployment through Optimizations in the
   Planning of ICT Networks   117
   R. Nielsen, A. Mihovska, O. Madsen and R. Prasad

10. Low Power Hardware Platforms   131
    A. Nannarelli

11. Alternative Power Supplies for Wireless and Sensor Networks
    and Their Applications   145
    N. Nakajima

12. Energy Saving in Wireless Access Networks   157
    L. Jorguseski, R. Litjens, J. Oostveen and H. Zhang

13. Optimizing Energy Usage in Private Households   185
    J. Rohde, S. Wolff, T.S. Toftegaard, P.G. Larsen, K. Lausdahl,
    A. Ribeiro and P.E. Rovsing

14. Photovoltaic-Aware Multi-Processor Scheduling – A Recipe
    for Research   211
    P. Koch

15. Quantum Information Technology for Power Minimum
    Info-Communications   229
    M. Sasaki, A. Waseda, M. Takeoka, M. Fujiwara and H. Tanaka

16. Approaches for Green Communication for Improving Energy
    Efficiency   245
    M. Umehira

17. Interconnect-Aware High-Level Design Methodologies for
    Low-Power VLSIs   265
    M. Kameyama and M. Hariyama

18. Development of Wideband Optical Packet Switch Technology   275
    H. Furukawa and N. Wada

Part III: Application and Business Models

19. Application of Green Radio to Maritime Coastal/Lake
    Communications and Locationing Introducing Intelligent
    WiMAX (I-WiMAX)   287
    X. Lian, H. Nikookar and L.P. Ligthart

20. A Futuristic Outlook on Emerging Business Models   309
    P. Lindgren, Y. Taran, K. Saughaug and S. Clemensen

Author Index   335

Subject Index   337

About the Editors   339

About the Authors   343

Preface

राजविद्या राजगुह्यं पवित्रमिदमुत्तमम् ।
प्रत्यक्षावगमं धर्म्यं सुसुखं कर्तुमव्ययम् ॥ २ ॥

raja-vidya raja-guhyam
pavitram idam uttamam
pratyaksavagamam dharmyam
su-sukham kartum avyayam

This knowledge is the king of education, the most secret of all secrets. It is the purest knowledge, and because it gives direct perception of the self by realization, it is the perfection of religion. It is everlasting, and it is joyfully performed.

The Bhagavad Gita (9.2)

This book is the outcome of the special session on Green Communications at "The 12th International Symposium on Wireless Personal Multimedia Communications" (WPMC'09) held in Sendai, Japan. The session was organised by Ramjee Prasad, CTIF-HQ, Denmark, Shingo Ohmori, CTIF-Japan, Dina Šimunić, University of Zagreb, Croatia. To the best of the editors' knowledge this is the first book on the Green Information and Communication Technologyies (ICT) covering broad perspective as illustrated in Figure 1.

ICT is playing an increasingly important role in both business and individual's private life. It has increased international interconnectedness and speed up the process of globalization. But on the other side the total energy consumption by the communication and networking devices and the relevant global CO_2 emission is increasing exponentially. ICT has, in many

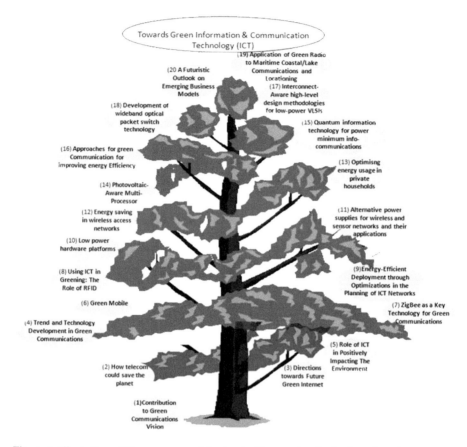

Figure 1 Illustration of the coverage of the book. The number in the branches denotes the chapter of the book.

ways, a vital role to play. It accounts for about two percent of global CO_2 emissions. Telecommunications applications can have a direct, tangible impact on lowering greenhouse gas emissions, power consumption, and achieve efficient recycling of equipment waste.

Implementing green practices is new ground for most ICT managers. The session on "Green Communications" has brought together academic and industrial researchers all around the world to discover "Green ICT" practices.

Within the framework of ICT, a number of paradigm-shifting technical approaches can be expected, including but not limited to energy-efficient network architecture & protocols, energy-efficient wireless transmission techniques, reduced transmission power & reduced radiation, cross-layer optimization methods, and opportunistic spectrum sharing without causing harmful interference pollution. This book has successfully addressed the various topics of research on Green ICT. The coverage of new and upcoming issues on Green ICT makes it a good choice for students, engineers and executives. We cannot claim that this book is errorless. Any remarks to improve the text and correct the errors would be highly appreciated.

Acknowledgements

The material in this book originates from the proceedings of the WPMC'09. Therefore, the editors would like to thank all the colleagues from the academia and industries involved in WPMC'09 for their collaboration and dedication that made the success of the conference and also helped to finalize this book. Special thanks to WPMC'09 General Chair, Professor Fumiyuki Adachi from Tohoku University, Japan.

We would also like to thank Dua Idris from CTIF for her support in completing the book project.

Credit for the major efforts in helping us to put the material together and shape it into a final version goes to Kirti Pasari from CTIF.

Abbreviations

2G	Second Generation
2zeroP	Towards Zero Power ICT
3G	Third Generation
3GPP	Third Generation Partnership Project
AC	Alternate Current
ADSL	Asymmetric Digital Subscriber Loop
Ah	Ampere hour
AIS	Automatic Identification System
AOFDM	Adaptive Orthogonal Frequency Division Multiplexed
AP	Access Point
AWGN	Additive White Gaussian Noise
BB/RF	Base Band / Radio Frequency
BER	Bit Error Rate
BERT	Bit Error Ratio Tester
BI	Beacon Interval
BM	Business Model
BPSK	Binary Phase Shift Keying
BS	Base Station
BSA	Best Sub carrier Allocation
BTSs	Base Transmission Stations
CAPEX	CAPital EXpenditure
CATV	Cable TV
CDMA	Code Division Multiple Access
CMA	Constant Modulus Adaptive
CMOS	Compementary Metal Oxide Semiconductor
CO_2	Carbon Dioxide
CO_2e	carbon dioxide equivalent
COP	Conference of the Parties
CPICH	Common Pilot CHannel

CSI	Channel State Information
CSSM	Coherent Signal Subspace Method
CTP	Cable Trench Problem
dB	decibel
DC	Direct Current
DDA	Decision Directed Algorithm
DFG	Data Flow Graph
DFT	Discrete Fourier Transform
DMFC	Direct Methanol Fuel Cell
DOA	Direction of Arrival
DoD	Depth of Discharge
DPD	Digital Pre-Distortion
DRX	Discontinuous Reception
DSP	Digital Signal Processing
DTIM	Delivery TIM
EC	European Comission
ECCP	European Climate Change Programme
EDGE	Enhanced Data Rates for GSM Evolution
EEC	Economic Integration Organization
EEE	Energy Efficiency Ethernet
EER	Energy Efficiency Rate
EER	Envelop Elimination and Restoration
EH	Energy Harvesting
EIA	American Energy Information
EIEEE	Environmental information on Electrical and Electronic Equipment
EPA	Environmental Protection Agency
EPON	Ethernet Passive Optical network
ESPRIT	Estimation Parameters via Rotation Invariance Technique
ET	Envelop Tracking
ETNO	European Telecommunication Network Operators' Association
ETS	Emissions Trading Scheme
EU	European Union
EUP	Energy Using Products
FDLs	Fiber Delay-Lines
FMBC	Fixed Mobile and Broadcasting Convergence
FP	Framework Programme
FPGA	Field Programmable Gate Array
FTTH	Fibre to the Home

FU	Functional Unit
GA	Genetic Algorithm
GaN	Gallium Nitride
GDP	Gross Domestic Product
GeSI	Global e-Sustainability Initiative
GHG	Green House Gas
GMPLS	Generalized Multi-Protocol Label Switching
G-PON	Gigabit PON
GPU	Graphical Processing Unit
GSM	Global System for Mobile communications
GSMA	GSM Association
Gt	billion tonnes
GWh	Giga Watt Hours
HDTV	High Definition TV
HEMT	High Electron Mobility Transistor
HGW	Home Gateway
HOTARU	Hybrid Optical Network Architecture
HPA	High Power amplifier
HSDPA	High Speed Downlink Packet Access
HS-DSCH	High Speed-Downlink Shared Channel
HSPA	High Speed Packet Access
I- WiMAX	Intelligent WiMAX
ICT	Information and Communication Technologies
ICT&CC	ICT & Climate Change
IEC	International Electrotechnical Commission
IF	Intermediate Frequency
IPC	Inter-Processor Communication
IPCC	International Panel on Climate Change
ISDN	Integrated Services Digital Networks
ISO	International Organization for Standardization
ITS	Intelligent Transport Systems
ITU	International Telecommunication Union
IXes	Internet eXchange points
kWh	Kilo Watt Hours
LD	Laser Diodes
LED	Light Emitting Diode
LEED	Leadership in Energy and Environmental Design
Li-ion	Lithium Ion
Li-Pol	Lithium Polymer

LMS	Least Mean Square
LN-IM	Linbo3 Intensity Modulator
LOS	Line of Sight
LTE	Long Term Evolution
MAC	Medium Access Control Layer
MEM	Maximum Entropy Method
MIMO	Multiple Input Multiple Output
ML	Maximum Likelihood
MLE	ML Estimator
MLLD	Mode-Locked Laser Diode
MLM	Maximum Likelihood Method
MMSE	Minimum Mean Square Error
MOE/MOD	Multiple Optical En/Decoders
MST	Minimum Spanning Tree
Mt CO2e	Million tonnes carbon dioxide equivalent
MUSIC	Multiple Signal Classifications
MVDR	Minimum Variance Distortion Response
MW-OPS	Multi-Wavelength Optical Packet Switching
NAV	Network Allocation Vector
NB	Network Bases
NC	Null Constraint
NGMN	Next Generation Mobile Networks
NiMH	Nickel Metal-Hybrid
OC	Optical Codes
OCS	Optical Circuit Switching
OECD	Organisation for Economic Co-operation and Development
OFDM	Orthogonal Frequency Division Multiplexed
OLT	Optical Line Terminal
ONUs	Optical Network Units
OPEX	OPerational EXpenditure
OPS	Optical Packet Switch
O-QPSK	Offset Quarternary Phase Shift Keying
OTDM	Optical Time-Division Multiplexed
PD	Photo Diodes
PDL	Polarization Dependent Loss
PEs	Processing Engines
PHY	Physical Layer
PoE	Power over Ethernet
PON	Passive Optical Network

PSD	Power Spectrum Density
PSK	Phase-Shift Keying
PSM	Power Saving Mode
PUE	Power Usage Effectiveness
PUE	Power Usage Effectiveness
PV	Photovoltaic
PVB	Partial Virtual Bitmap
QAM	Quadrature Amplitude Modulation
QoS	Quality of Service
QPSK	Quadrature Phase Shift Keying
RAN3	Radio Access Network Group 3
RAT	Radio Access Technology
RF	Radio Frequency
RFID	Radio Frequency Identification
RoHS	Reduction of Hazardous Substances
SA	Smart Antenna
SAE	System Architecture Evolution
SBU	Strategic Business Unit
SC	Single Carrier
SCORE	Spectrum self COherent REstoral
SDR	Software Defined Radios
SFQ	Single Flux Quantum
SINR	Signal to Interference and Noise Ratio
SMA	SubMiniature version A
SMI	Sample Matrix Inversion
SMT	Steiner Minimal Tree
SNR	Noise Ratio
SoC	Systems on Chip
SOCRATES	Self-Optimisation & self-ConfiguRATion in wirelEss networkS
SOFDMA	Scalable Orthogonal Frequency Division Multiplexed
SPST	Shortest Path Spanning Trees
SPU	Strategy and Policy Unit
SQoS	Structural Quality of Service
SR	Smart Radio
SS	Subscriber Station
STB	Set-Top Box
STI	Sleep Transistor Insertion
TC	Technical Committee
TCAM	Ternary Content Addressable Memory

TDM	Time Division Multiplexing
TDMA	Time Division Multiple Access
TOA	Time of Arrival
TSAG	Telecommunication Standardization Advisory Group
TSB	Telecommunication Standardization Bureau
UCS	Universal Charging Solution
UI	User Interface
UMTS	Universal Mobile Telecommunication System
UNCED	United Nations Conference on Environment and Development
UNEP	United Nations Environment Programme
UNFCCC	United Nations Framework Convention on Climate Change
VGND	Virtual Ground Line
VHF	Very High Frequency
VoD	Video on Demand
VOIP	Voice over IP
WAVES	Weighted Average of Signal Subspace
WCED	World Commission on Environment and Development
WDM	Wavelength Division Multiplexing
WEEE	Waste Electrical and Electronic Equipment
Wh	Watt hour
WHO	World Health Organization
WiBro	Wireless Broadband
WiFi	Wireless Fidelity
WiMAX	Worldwide Interoperability for Microwave Access
WLAN	Wireless local Area Network
WSNs	Wireless Sensor Nodes
WSSD	Summit on Sustainable Development
WWF	World Wide Fund for Nature
ZXD	Zero-Crossing Demodulation

PART I

OVERVIEW

1

Contribution to Green Communications Vision

Dina Šimunić[1], Ramjee Prasad[2] and Shingo Ohmori[3]

[1]*Faculty of Electrical Engineering and Computing, University of Zagreb,
10000 Zagreb, Croatia*
[2]*Department of Electronic Systems, Aalborg University, 9220 Aalborg, Denmark*
[3]*CTIF-Japan, Yokosuka Research Park (YRP) Center No. 1 Bldg., 3-4 Hikarinooka,
Yokosuka, Kanagawa 239-0847, Japan*

Wouldn't it be wonderful to live in a green
world, served by the full power technology and
thus living at the highest possible standards?

Abstract

The WPMC 2009, held in Sendai, Japan, is one of the first conferences
which brought together four sessions on green communications, bringing
together academic and industrial researchers for discussing energy-efficient
communications. It covered very broad range of topics from energy efficiency for IT to wireless sensor networks on the other side. Information and
communication technologies are mature to provide support to many other in-
dustrial/governmental/public sectors in the on-going "information era" of the
2nd Millenium. The information and communication technologies provide
an energetic road, which does not connect only two points, but is open to
the global approach and gives possibility to everyone to be involved in the
road at every moment. At this very moment, the best ICT potential in our
everyday life has still not yet been properly explored. Since the information
and communication technologies are the foundation for the holistic approach

R. Prasad et al. (Eds.), Towards Green ICT, 1–28.
© 2010 *River Publishers. All rights reserved.*

in the "green world", we expect it to be done in the very near future. Our chapter and this book are a small contribution toward this great perspective.

Keywords: Green communications, green world, smart ICT, smart energy, smart housing, dematerialization, smart transport.

1.1 Introduction

The World Commission on Environment and Development, sponsored by the United Nations, conducted a study of the world's resources and how these are being depleted and published the 1987 report entitled "Our Common Future", also known as the Brundtland Commission Report. The term "sustainable development" was introduced and defined as "development that meets the needs of the present without compromising the ability of future generations to meet their own needs" [1]. A more precise definition of sustainable development would be an enlargement of the preceding, and is given as "the design of human and industrial systems to ensure that humankind's use of natural resources and cycles do not lead to diminished quality of life due either to losses in future economic opportunities or to adverse impacts on social conditions, human health, and the environment" [2]. These requirements reflect that the interaction of social conditions, economic opportunity, and environmental quality is essential for making a next step in human development, encompassing fundamental research, education and knowledge transfer, and leading to the new paradigm of green engineering.

The United Nations Conference on Environment and Development (UNCED), or the Earth Summit, held in Rio de Janeiro in 1992, communicated the idea that sustainable development is both a scientific concept and a philosophical ideal. A resultant document, "Agenda 21", was endorsed by 178 governments and hailed as a blueprint for sustainable development. Ten years later, in 2002, the World Summit on Sustainable Development (WSSD) identified five key major areas for moving sustainable development plans forward: water, energy, health, agriculture, and biodiversity [3].

In 2007, the Nobel Peace Prize was awarded jointly to former US vice-president Al Gore and to the United Nations Intergovernmental Panel on Climate Change (IPCC) with a citation "for their efforts to build up and disseminate greater knowledge about man-made climate change, and to lay the foundations for the measures that are needed to counteract such change" by the Norwegian Nobel Committee [4].

As of 6 November 2009, 189 countries and one regional economic integration organization (the EEC) have deposited instruments of ratification, accession, approval or acceptance of the legally binding Kyoto Protocol, the agreement negotiated via the United Nations Framework Convention on Climate Change (UNFCCC) [5]. The Kyoto protocol sets a target for average global carbon emissions reductions of 5.4% relative to 1990 levels by 2012, with ongoing discussions for a post-2012 period. Particular targets have been developed by individual regions and countries. In 2007, the European Union (EU) announced a 20% emissions-reduction target compared to 1990 levels by 2020, and will increase this to 30% with an international agreement in the post-2012 period. The UK is aiming for a reduction of 60% below 1990 levels by 2050, with an interim target of about 30%. Germany is aiming for a 40% cut below 1990 levels by 2020, while Norway will become carbon-neutral by 2050. The climate change legislation of California (AB 32), commits the state to 80% reductions below 1990 levels by 2050. China's latest five-year plan (2006–2010) contains 20% energy efficiency improvement targets, with allowance for reduction of the impact of recent fuel shortages on its economic growth.

The European Commission released "Communication from the Commission to the European Parliament, the Council, the European Economic and Social Committee and the Committee of the Regions: Towards a Comprehensive Climate Agreement in Copenhagen" [6]. The position paper "addresses three key challenges: targets and actions; financing of low-carbon development and adaptation; and building an effective global carbon market". The European Union (EU) had committed to implementing binding legislation, even without a satisfactory deal in Copenhagen and thus became the world leader for the post-2012 period. The most important part is in the revision of its carbon allowances system called the Emissions Trading Scheme (ETS) designed for the post-Kyoto period (after 2013), initially founded in January 2005, based on Directive 2003/87/EC, which entered into force on 25 October 2003 [7]. The Directive has been amended by Directives 2004/101/EC of 27 October 2004 [8]; 2008/101/EC of 19 November 2008 [9], EC Regulation No. 219/2009 of 11 March 2009 [10] and Directive 2009/29/EC of 23 April 2009 [11]. The EU Commission will foresee that sectors exposed to international competition should be granted some free allocations of CO_2 emissions, whereas other sectors should buy such credits on an international market. Energy intensive industries in Europe have advocated for this benchmark system in order to keep funds in investment capacities for low carbon products rather than for speculations.

The 2009 United Nations Climate Change Conference, known as Copenhagen Summit was held in Copenhagen, Denmark, 7–18 December 2009. It included the 15th Conference of the Parties (COP) to the United Nations Framework Convention on Climate Change and the 5th Meeting of the Parties (COP/MOP5) to the Kyoto Protocol. The media (e.g., Anne Eckstein, Europolitics, Tuesday 22 December 2009, [12]) reported about dissapointment of EU-27 with the outcome of the Copenhagen Summit. Furthermore, The Copenhagen Accord [13] was drafted by the United States (US), China, India, Brazil and South Africa on December 18, and judged a "meaningful agreement" by the US government. The next day, the document did not pass unanimously by all the participating countries and thus it is not a legally binding treaty. The document underlines that climate change is one of the greatest challenges of the present and that actions should be taken to keep any global temperature increase to below 2°C.

1.2 Role of ICT

In all key areas, information and communication technologies (ICT) are seen as the number one innovation engine. As an example, ICT drives more than 80% of innovations in Germany's strong application sectors, such as automotive, medical technology and logistics [14]. Thus, a key path to sustainable development and reduction of carbon emissions, which should lead to healthier life on the planet Earth, is to introduce and increase development of ICT for all five key areas mentioned above, but more specifically to at least the following five proposed areas of human activity: smart grids, smart housing, dematerialization, smart transport (including smart motor systems), and smart environment. The introduction of ICT to these areas will significantly contribute to reductions of man-made green-house gas (GHG) emissions.

The growth of ICT also means positive contributions to global GDP growth and globalization. A third of the economic growth in the Organisation for Economic Cooperation and Development (OECD) countries between 1970 and 1990 was due to access to fixed-line telecoms networks alone, with associated lowered transaction costs. In low income countries, an average of 10 more mobile phone users per 100 people was found to stimulate a per capita GDP growth of 0.59%. As an example, improved communication in China has helped increase wealth by driving down commodity prices, coordinating markets and improving business efficiency. Globally, the ICT sector contributed 16% of GDP growth from 2002 to 2007, and the sector itself has increased its share of GDP worldwide from 5.8 to 7.3%. The ICT sector's share of the economy is predicted to jump further to 8.7% of GDP growth worldwide from 2007 to 2020.

All human beings have a moral imperative for a healthy environment, evidenced as environmental or green conscientiousness. For engineers all over the world the 2005 statement by the United States National Academy of Engineering that "engineers will continue to be leaders in the movement toward the use of wise, informed, and economical sustainable development", remains of tremendous importance [15]. It is imperative that the term "green world", defining sustainable designs for all devices, becomes an integral part of every engineering teacher and disciple. The term "green engineering" is defined by the US Environmental Protection Agency (EPA) as "the design, commercialization and use of processes and products that are feasible and economical while reducing the generation of pollution at the source and min-imizing the risk to human health and environment" [16]. In summary, green engineering can be defined as environmentally conscientious attitudes, values and principles, combined with science, technology, and engineering practice, all directed towards improving local and global environmental quality. The very important property of green engineering is that decisions to protect human health and the environment are planned according to the basic engin-eering principles of "foresight" thinking, and as such can have the greatest impact when applied early to the design and development phase of a process or product.

The nine main principles of green engineering were given as conclu-sions of the conference "Green Engineering: Defining the Principles" held in Sandestin, Florida, USA, in May 2003 [17]. These principles are:

1. holistic processing and integration of environmental impact assessment tools;
2. conservation and improvement of natural ecosystems while protecting human health and well-being;
3. using life-cycle thinking in all engineering activities;
4. ensuring that all material and energy inputs and outputs are as inherently safe and benign as possible;
5. minimization of depletion of natural resources;
6. prevention of waste as much as possible;
7. development and application of engineering solutions, while being cognizant of local geography, aspirations and cultures;
8. improvement and creation of new and innovative technologies to achieve sustainability; and

9. active engagement of communities and stakeholders in development of engineering solutions.

Of course, the highest importance is introduction of ICT resources into the domain of climate change. This can be achieved via monitoring climate change, via mitigation of local effects of climate change, via concerted action against global warming and via ICT standardization in the field of climate change. However, one should keep in mind that aside from emissions associated with deforestation, the largest contribution to man-made GHG emissions comes from power generation and fuel used for transportation. Therefore it is not surprising that the biggest role ICTs could play is in helping to improve energy efficiency in power transmission and distribution, in buildings and factories that demand power and in the use of transportation to deliver goods. ICT could play a significant role in mitigating global carbon emissions from motor systems and industrial process optimization, up to 970 MtCO2e in 2020.

For instance, motor systems generally currently use 70% of total industry electricity consumption. In Eastern countries, they are cca 20% less energy efficient than those in Western countries. By 2020, industrial motor systems in China will be responsible for 34% of power consumption and 10% of carbon emissions, or 1 to 2% of global emissions. USA investigations discovered that 45% of all vehicle miles driven in the world are driven in the USA. Thus, it would be great if, for instance, the USA and China would make especial efforts related to improvement of smart transport and dematerialization. As an example from the USA, "Motor Decisions Matter" [18] gives a set of ongoing policies and practices that help commercial and industrial facilities effectively manage their motor populations, thus decreasing energy costs and improving productivity. The Australian "Energy Smart Business Program" [19] states that properly sized, energy efficient motors with electronic variable speed drives and improved gears, belts, bearings and lubricants use only 40% as much energy as standard systems and, in financial terms, with a four-year payback project, variable speed drives installations for the control of conveyors and combustion and ventilation fans can deliver energy savings upwards of AUS$120 million (EUR 73 million) a year. The Motor Challenge Programme, a voluntary programme, has been promoted by the European Commission to help companies improve the energy efficiency of their electric motor driven systems. The programme has reached conclusions that switching to energy efficient motor driven systems would help to save Europe up to 202 billion kWh in electricity consumption, equivalent to a reduction of EUR 10 billion per year in operating costs for industry, as well as a reduction of 79 million tonne of CO_2 emissions (calculated only in EU 15), or approximately a quarter of the EU's Kyoto target. This reduction is the annual amount of CO_2 that a forest the size of Finland transforms into oxygen. If industry is allowed to trade these emission reductions based on energy saved, this would generate a revenue stream of EUR 2 billion per year. Calculation for EU 25 gives even higher reduction potential of 100 million tonne. Finally, it would mean a 6% reduction in Europe's energy imports (EU 25) [20].

All of these facts contribute to the idea that solving the global issues should be performed on the global level, because knowledge, ideas, human, engineering and economic resources are distributed, and only by a joint action can the GHG issue be solved in the best way for our planet, as well as for the global economy.

Here one should add the very important role of standardization. The role of International Electrotechnical Commission (IEC) Technical Committee (TC) 111: "Environmental standardization for electrical and electronic products and systems, and implementation of standards in the regions" [21] is crucial in the field of green engineering. The new Chairman of IEC TC 111, Yoshiaki Ichikawa of Hitachi Ltd., Japan, noted that three major international standardization bodies: the IEC, ISO (International Organization for Standardization) and ITU (International Telecommunication Union), have committed themselves to avoid duplication of work, which nowadays is an issue even for terminology, which is not unified across the three organizations. Dr. Ichikawa also noted that "there are initiatives in environmental legislations that have been mirrored in Japan, who is a leader in this respect." Of course, legislation issues have been discussed in the USA and the European Commission. IEC TC 111 works closely with the IEC Standardization Management Board Advisory Committee on Environmental Aspects (SMB ACEA).

Standardization of environmental aspects concerns: preparation of the necessary guidelines, basic and horizontal standards, including technical reports, in the environmental area, in close cooperation with product committees of IEC, which remain autonomous in dealing with the environmental aspects relevant to their products; liaison with product committees in the elaboration of environmental requirements of product standards in order to foster common technical approaches and solutions for similar problems and thus assure consistency in IEC standards; liaison with ACEA and ISO/TC 207 on Environmental Management; and monitoring closely the corresponding regional standardization activities worldwide in order to become a focal point for discussions concerning standardization.

Publications issued by IEC TC 111 are: IEC 62321: "Electrotechnical products – Determination of levels of six regulated substances (lead, mercury, cadmium, hexa valent chromium, polybrominated biphenyls, polybrominated diphenyl ethers)"; IEC 62430: "Environmentally conscious design for electrical and electronic products"; IEC/PAS 62545: "Environmental information on Electrical and Electronic Equipment (EIEEE)" and IEC/PAS 62596: "Electrotechnical products – Determination of restricted substances – Sampling procedure – Guidelines".

ITU is involved in the protection of the environment by the role of telecommunications and information technologies. It started at the Plenipotentiary Conference, 1994 (Resolution 35, Kyoto) and at the World Telecommunication Development Conferences, in 1998 (Resolution 8, Valletta), in 2002 (Recommendation 7, Istanbul) and in 2006 (Resolution 54, Doha). In 2007, ITU and its membership and partners launched a major programme to investigate the specific relationship between ICTs and climate change. Technology Watch Briefing Report was reviewed at the meeting of the Telecommunication Standardization Advisory Group (TSAG) in December 2007. TSAG provided advice to the Director of the Telecommunication Standardization Bureau (TSB) on a number of actions, including the holding of two symposia during the first half of 2008 (the first in Kyoto, Japan, 15–16 April 2008, co-organized and hosted by the Ministry of International Affairs and Communications (MIC); the second in London, UK, 17–18 June 2008, supported and hosted by BT plc) [22].

As a conclusion, any mitigation strategy must have transdisciplinary elements and ICTs can help with this, either in a direct way, by reducing the ICT sector's own energy requirements (e.g., Next-Generation Networks are expected to reduce energy consumption by 40% compared to today's PSTN), or in an indirect way, through using ICTs for carbon displacement (e.g., joint initiative of European Telecommunication Network Operators' association, ETNO, and the World Wide Fund for nature, WWF, [23]). Also, ICT can provide the technology to implement and monitor carbon reductions in other sectors (e.g., intelligent transport systems, ITS) and thus would help reduce energy consumption in a systemic way. Therefore, a key way to sustainable development and reduction of carbon emissions, which should lead to healthier life on the planet Earth is to introduce and more develop ICT to the five key areas via indirect or systemic way: smart grids, smart housing, dematerialization, smart transport, including smart motor systems, and smart environment. The introduction of ICT to these areas will significantly contribute to reduction of man-made Green House Gas (GHG) emissions.

1.2.1 Smart Energy and Smart Grids

Energy is everywhere in our surrounding environment, available in the various forms of mechanical, thermal or geothermal, electromagnetic (including light), natural, human body or other energy. Mechanical energy derives from sources such as vibration, mechanical stress and strain, thermal energy from waste energy from furnaces, heaters, and friction sources, or as geothermal energy from Earth, electromagnetic from inductors, coils and transformers, light from sunlight or room light, natural energy from the environment such as wind, water flow, or ocean currents. Human body energy is a combination of mechanical and thermal energy naturally generated or through physical activities actions such as walking or jumping. Other energy comes from chemical and biological sources. It is important to note that all the energy sources mentioned are virtually unlimited and free of charge, if they can be captured at or near the system location. However, due to the inadequate energy conversion devices, as well as sometimes inadequate quantities found in the immediate surroundings, the energy from these sources cannot supply adequate power for any viable purpose. In fact, until recently it has not been possible to capture such energy sufficiently to perform any useful work. The information era with ICT as enabling technologies is about to change the existing energy scenario.

Nowadays, energy is at the forefront of the global political and social agenda. It is of crucial importance to harness energy in the most efficient and sustainable way using comprehensible and comparable quantities. The IPCC together with the World Health Organization (WHO) and the United Nations Environment Programme (UNEP) showed the human-induced increase in concentration of three important gases, viz. carbon dioxide, nitrous oxide, and methane, which is causing global warming. It is shown that the concentration of these in the atmosphere has been constantly increasing since the year 1800. As a matter of fact, the concentration of carbon dioxide has risen by more than 30%. The majority of climate scientists agree that a further increase in the gases will cause a rise in the Earth's temperature. Indicators of the human influence on the atmosphere during the Industrial Era are given in Figure 1.1 [24].

According to the United Nations Framework Convention on Climate Change (UNFCCC), large scale mitigation actions in terms of "combating cli-

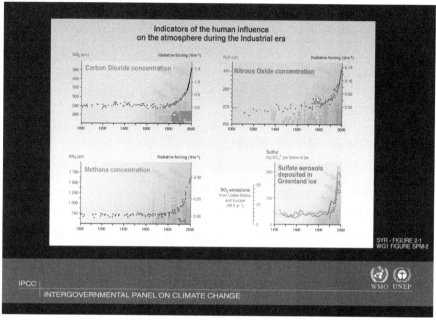

Figure 1.1 Indicators of the human influence on the atmosphere during the Industrial Era.

mate change and the adverse effects thereof" are needed mostly by developed countries. The cost of mitigation by the year 2030 has been calculated, and it would not exceed 3% of the global GDP for limitation of average temperature increase to 2.0 to 2.4°C. Along with the temperature increase limitation goes many benefits, such as lower air pollution and associated health benefits, higher energy security, larger employment and stable agricultural production, meaning also greater food security. For a temperature increase of 2.0°C, the sea-level could rise from 0.4 to 1.4 meters on account of thermal expansion alone, which together with ice-melting across the globe would lead to the submerging of several small island states and Bangladesh. Thus, it seems reasonable to try to limit temperature increase to 2.0°C by means of having a definitive peak of global emissions no later than the year 2015 [25].

Another group of initiatives targets so-called "smart grid". The idea of the smart grid is to provide real-time information to end-users, grid operators and distributed generators. In this way everybody can accomplish the tasks with minimized emissions for the benefit of all. This can be accomplished by changing the grid network from point-to-node to point-to-point with the connection between end-users, by process automation using smart

ICT-meters, and by creating a real-time marketplace for energy consumption by enabling two-way communication between customers and suppliers. Thus, energy saving in residential network has been already developed and applied in some countries (e.g. in the USA "net metering" is used as an alternative to battery storage, which makes use of the observation that electricity and money are routinely converted one into the other, which further on encourages and educates users for energy/money saving) [26]. The European Union is encouraging research in the field of application of ICT for energy management, because communication infrastructures are seen as a key for monitoring and trading energy in oil/gas/electrical energy networks. One should note that reduction of losses in India's power sector by 30% would be possible solely through better monitoring and management of electricity grids, first with smart meters and then by integrating more advanced ICTs into the so-called energy internet. Smart grid technologies could globally reduce 2.03 GtCO$_2$e, worth EUR 79 billion. The ICT is a foundation for implementation of electric mobility, for supply chains etc. The smart grids on a large scale are seen as encompassing a concept of green communications through intelligent distributed sensor and actuator networks, with corresponding backbone connection. The Internet of Things [27] can thus help improve future smart energy provisioning networks, in which units for energy production and consumption can be managed in a flexible way. Of course, for achieving such an ambitious goal, the joint research and development activities with energy network and supply chain management are required to elaborate sustainable architectures for an integrated system approach.

Another positive example is the one of the Ministry of Internal Affairs and Communications (MIC) of Japan, which has developed a handbook for corporations and organizations that use ICT systems with the aim of providing guidelines and advice toward limiting the negative impacts while enhancing the positive impacts on the environment when ICT systems are introduced, operated, and disposed [28]. The outlines of the handbook were given as a contribution to the final report of the ITU-T Focus Group on ICT & Climate Change (ICT&CC), established by TSAG in July 2008, and which completed its work in April 2009 especially with Deliverable 4 – Direct and indirect impact of ITU-T standards.

1.2.2 Smart Housing

Smart housing or smart buildings relate to the design and technology which would enable buildings with minimal, or even negative, energy consumption.

This especially concerns buildings of public importance, such as hospitals and schools.

In the building area, the Leadership in Energy and Environmental Design (LEED) [29] certification is a standard which many new building projects strive to achieve. This means the building uses a substantial amount of green engineering, which also comprises a significant amount of ICT. Namely, ICT present new opportunities for monitoring the physical environment, further increasing the complexity of interaction of in-built electronic and physical entities. The ICT can be used for monitoring, pricing and control of physical flows and spaces in cities. In this way, the environmental information systems carrying data about a range of air pollutants, noise levels and water pollutants at different spatial scales, remotely and in real-time, could be given for public access.

Even more effective monitoring and management, followed by substantial energy bill savings, can be performed via smart metering technology in the resource flows along the networks (e.g. energy and water) [30]. As already mentioned for Netmetering, physical resources flows can be charged or re-charged, according to changing levels of demand (peak or non-peak level). This energy-to-money conversion is very educational for the general public to set targets for reductions in energy and water consumption, and as such smoothes out peak demands, enabling reduction of expensive investment in infrastructure such as roads, power stations and waste treatment plants. Except at the level of resource flows, ICT proves to be very effective in energy reduction by means of optimization of quality of ventilation, lighting and security control systems, and these can be applied either individually or at the network level. For instance, better building design, management and automation could save 15% of North America's buildings emissions, and global estimation is for 1.68 $GtCO_2e$ of emissions savings, worth EUR 216 billions.

1.2.3 Dematerialization

The topic of smart living relates among others to the very important topic of travel substitution (or so-called dematerialization, related to replacement of "atoms" with "bits") to savings in energy consumption due to flexible places of work, and with the possibility for virtual meetings, which is already definitely possible for number of jobs due to the present existing level of development of ICT.

As is very well known, ICT already has a great role in human development, and with incredible future capabilities for expansion. For instance, ICT contributed to both the centralization and decentralization of cities. Centralization occured with the development of skyscrapers in the early part of the 20th century, which would not have been possible without the telephone, ensuring communication within multi-storey towers. Decentralization occured later, when public transport facilitated movement to the suburbs, where again communication by the telephone facilitated urban living habits. These trends have continued to the present day, and the emerging ICT infrastructure allows even higher dispersion from cities to new locations in rural areas. The increasing separation between work and home, or leisure, recreation and shopping and home has a two-fold effect for energy consumption. On one hand, ICT flows replace physical distance, allowing more flexible life and saving energy; on the other hand, the trips are longer, requiring more energy and waste of infrastructure. This is exactly where the ICT flows are or will be used to save transport and travel energy consumption, by providing efficient transport and managing all kinds of travel networks. Figure 1.2 shows the impact of dematerialization. If significant number of people work from home more than three days a week, this could lead to energy savings of 20–50%, even with the increase in energy used at home or non-commuter travel. However, one should keep in mind that the impact of working from home varies depending on the amount of time spent at home, and the efficiency of the economy in which teleworking is introduced. Namely, it depends on the number of days per week the home-working is taking place, and if the take-up is lower than three days a week then the both working space (the office and home) should be maintained. Also, it depends on the transportation means of the teleworker. Finally, in efficient countries, such as Japan, the impact of teleworking may be reduced. One estimation is that dematerialization could be responsible for reducing emissions by 500 $MtCO_2e$.

1.2.4 Smart Transport

Road transportation efficiency improvement opportunities concern optimization of individual and commercial logistics. The oft-used term describing this area is Intelligent Transport Systems (ITS) [31]. ITS involves the applied use of various engineering disciplines, specifically ICT, enabling technologies and management strategies to transport infrastructure and to facilitate modern travel and transport operations as well as policy development, and in order to deliver increased efficiency, safety and informed mobility of people and goods. Probably, the the traffic light can be considered to be the first ITS device. But the incredible ICT growth and the rate of change in the

GtCO₂e

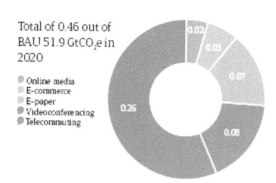

Total of 0.46 out of
BAU 51.9 GtCO₂e in
2020

- Online media
- E-commerce
- E-paper
- Videoconferencing
- Telecommuting

0.02
0.03
0.07
0.08
0.26

Figure 1.2 The impact of dematerialization. Source: *Expert Interviews*, January–March 2008.

technological environment, a rate which shows no sign of decreasing, has already facilitated many and still yet unknown applications in the transport sector, e.g. improved road safety, traffic management systems, and the provision of information to and from vehicles via navigational aids and/or vehicle performance. ITS also enables seamless and efficient financial transactions, reflected in less congested tolls stations thus enabling higher capacity. ITS is important in terms of security, because it facilitates behavioural monitoring via speed cameras, and increased safety through preventive (e.g. collision avoidance or speed limitation) and responsive measures (e.g. EU eCall [32], etc.). ITS has a huge environmental impact, because of the increased transparency of transport operations through asset (i.e. vehicle) tracking in fleet management systems. In conclusion, via use of ICT systems, ITS increases mobility, efficiency, attractiveness and capacity of existing physical infrastructure, thus enabling huge energy savings and decreasing the rate of thermal global increase.

With improved efficiency in transport and storage, in Europe only savings of fuel, electricity and heating could be 225 MtCO₂e. It is estimated that the global emissions savings in 2020 could be 1.52 GtCO₂e, with energy savings worth EUR 280 billion. As mentioned already, manufacturing in China without optimization would lead to 10% of China's emissions (2% of global emissions) in 2020. This will come solely from China's motor systems and to improve industrial efficiency even by 10% would deliver up to 200 million tonnes (Mt) CO₂e savings. Applied globally, optimized motors and industrial automation would reduce 0.97 GtCO₂e in 2020, worth EUR 68 billion.

Smart Environment

The ICT industry plays a significant role in setting an example for other industries and encouraging the adoption of ICT in all the areas, but especially in the field of its involvement in reduction or mitigation of climate change. It can further drive development and penetration of new technologies by enhancing the user-experience and reducing costs.

At level of urban environmental policy, environmental monitoring technologies have now started to raise fundamental issues for conceiving the city as both electronic and physical space. In this sense, a role of ICT in environmental policy is inevitable in its complexity [33].

ICT could be applied not only as monitoring tool, but also as an active tool for smart environment (for example, in energy harvesting and local storage, which are key points in energy research agenda). The both: energy harvesting and local storage, require development of radically new strategies. Energy harvesting is the process of efficient and effective capturing minimum amounts of energy from one or more of naturally-occurring energy sources. Examples of energy forms are mechanical, thermal, electromagnetic (including light), natural, human body or any other (e.g. biological or chemical). Local storage takes care of accumulating energy and storing it for later use. The aim of an energy harvesting system is to perform all these functions to power a variety of sensor and control circuitry for intermittent duty applications. In this way, energy harvesting, i.e. free energy source sensor is available throughout its lifetime maintenance-free. As an example, a ZigBee-enabled wireless sensor network can greatly benefit by using energy harvested power in an application of a remote wireless node without a wall plug and available battery. This gives remote control node a possibility for a self-powered electronic system, and a total independence in terms of need for its energy recharge. Of course, energy harvesting can be also used as an alternative energy source to supplement a primary power source and to enhance the reliability of the overall system and prevent power interruptions [34].

Lord Stern, author of the Stern Review, former UK Government and World Bank Chief Economist [35], points out that ignoring rising carbon emissions now will result not only in dangerous climate change in the future, but also they will be equivalent to losing at least 5% of global gross domestic product (GDP) each year, which would give estimates of damage to 20% of global GDP or more. On the other hand, the costs of reduction of GHG emissions to avoid the worst impacts of climate change can be limited to around 1% of global GDP each year.

1.3 Smart ICT

In 1965 Gordon Moore observed that the density of transistors in integrated circuits was doubling every 18 months. Now famously known as Moore's Law, this phenomenon has continued to the present day and has meant that the energy consumption per bit of information processed or transmitted has fallen by many orders of magnitude [36].

However, absolute growth in the use of digital technologies in developed world economies has led to an ever increasing carbon footprint. Thus, without major paradigm shifts in technological development, the growth in both usage and footprint is likely to continue as more and more people worldwide enter the digital age.

The ICT sector's own emissions are expected to increase, in a business as usual (BAU) scenario, from 0.53 billion tonnes (Gt) carbon dioxide equivalent (CO_2e) in 2002 to 1.43 $GtCO_2$e in 2020. However, specific ICT opportunities can lead to emission reductions five times the size of the sector's own footprint, up to 78 $GtCO_2$e, or 15% of total BAU emissions by 2020. Thus, the direct impact of ICT products and services and its enabling role in climate change solutions can be estimated by answering three main questions: what is the direct carbon footprint of the ICT sector; what are the quantifiable emissions reductions that can be enabled through ICT applications in other sectors of the economy and what are the new market opportunities for ICT and other sectors associated with realising these reductions.

The proposal of "SMART 2020: Enabling the low carbon economy in the information age" [37] is that ICT sector can contribute to emission reductions in a number of "SMART" ways: via **S**tandardization (by provision of information in standard forms on energy consumption and emissions, across sectors), **M**onitoring (in the design and control for energy use), **A**ccountability improvement (energy and carbon), **R**ethinking (innovations in energy efficiency opportunities across buildings/homes, transport, power, manufacturing and other infrastructure and provide alternatives to current ways of operating, learning, living, working and travelling) and **T**ransforming (smart and integrated approaches to energy management of systems and processes, including benefits from both automation and behaviour change and develop alternatives to high carbon activities, across all sectors of the economy).

In 2007, the total footprint of the ICT sector was 830 $MtCO_2$e, which is about 2% of the estimated total emissions from human activity released that year. This includes three main areas: personal computers (PC) and peripherals, telecoms networks and devices and data centres. In the "information era"

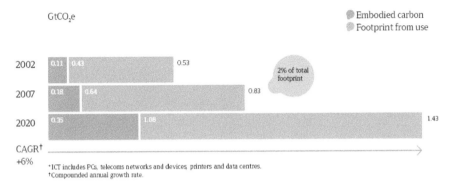

Figure 1.3 The global ICT footprint.

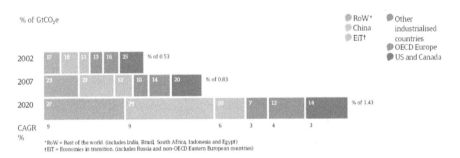

Figure 1.4 The global ICT footprint by geography.

of the 21st century, the percentage will grow at 6% each year until 2020. The carbon generated from materials and manufacture is about one quarter of the overall ICT footprint, the rest coming from its use (Figure 1.3) [37].

Figure 1.4 shows that increasing demand for ICT will give the most significant growth of ICT footprint in developing countries [37]. The developing countries (e.g. China) will have a need for more ICT devices and by 2020 they will catch up with developed countries, meaning that their ICT's carbon emissions will be 60% of the global (compared to cca 30% today). These numbers will be largely driven by growth in mobile networks and PCs. But, the fastest-growing elements of the footprint are data centres due to the need for storage and computing [37], which is shown in Figure 1.5.

The PCs' and monitors' combined carbon footprint was 200 MtCO$_2$e in 2002 and estimations show that with the a growth rate of 5% per annum it will be tripled by 2020 to 600 MtCO$_2$e. The global data centre footprint was 76 MtCO$_2$e in 2002. This number is expected to be more than triple by 2020 (259

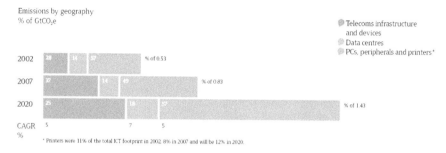

Figure 1.5 The global footprint by subsector.

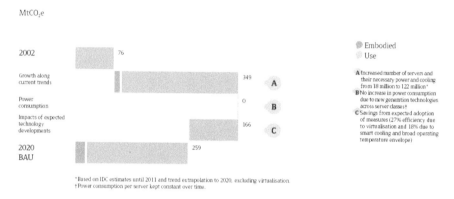

Figure 1.6 The global data centre footprint.

MtCO$_2$e), confirming to be the fastest-growing contributor to the ICT sector's carbon footprint, at 7% per year in relative terms, shown in Figure 1.6 [38].

The question on the global increase of power consumption definitely relates ICT domain in the sense of waste of important part of power in data/telecommunications centres, where chillers are sending cold air through floor vents into ICT equipment racks for their cooling and warm/hot air vents out the back of the recks. The air rises and recirculates back, of course, with an added energy for cooling. The US EPA provided the data on energy costs of data centers, where it has been shown that in six years the energy costs have been more than doubled (from 2000 to 2006 the energy costs increased from 28 to 62 TWh per year).

With regard to ICT equipment, not only data centres are energy sinks. Nowadays, every home has some ICT equipment with a tendency to expand. If one could limit energy sinks in every home on Earth, a considerable amount of energy could be saved. The European Commission is estimating that by

Global telecoms emissions %

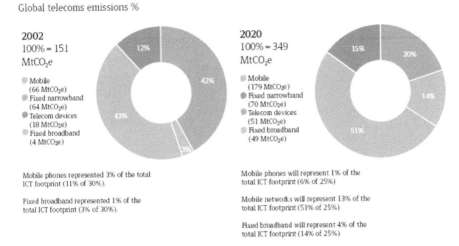

2002
100% = 151
MtCO₂e

- Mobile
  (66 MtCO₂e)
- Fixed narrowband
  (64 MtCO₂e)
- Telecom devices
  (18 MtCO₂e)
- Fixed broadband
  (4 MtCO₂e)

Mobile phones represented 3% of the total
ICT footprint (11% of 30%).

Fixed broadband represented 1% of the
total ICT footprint (3% of 30%).

2020
100% = 349
MtCO₂e

- Mobile
  (179 MtCO₂e)
- Fixed narrowband
  (70 MtCO₂e)
- Telecom devices
  (51 MtCO₂e)
- Fixed broadband
  (49 MtCO₂e)

Mobile phones will represent 1% of the
total ICT footprint (6% of 25%)

Mobile networks will represent 13% of the
total ICT footprint (51% of 25%)

Fixed broadband will represent 4% of the
total ICT footprint (14% of 25%)

Figure 1.7 Global telecoms footprint (devices and infrastructure).

the year 2012, the energy consumption in a home will reach 50 TWh, from 0
TWh in the year 2000 [39]. Actually, the estimation is that of the 10% overall
global energy spent for ICT, today its 70% are being spent in homes/offices
and only 30% in the network/server farms. Data rates in wired and wireless
networks are driven by Moore's Law and are thus rising by a factor of 10
every 5 years.

Figure 1.7 shows the global situation in telecoms infrastructure and
devices, where a parallel increase in fixed and mobile occurred. It is expected
that fixed-line, narrowband and voice accounts remain fairly constant. However, the number of broadband accounts, in both telecom segments will grow
to more than double in the period from 2007 to 2020. Also, in the same period
it is expected that the number of mobile accounts will come to almost double
number. In the five years period, from 2002 to 2007, the growth in telecoms
emissions has doubled (from 150 MtCO₂e in 2002 to 300 MtCO₂e in 2007).
With implementation of new technologies, it is expected to reach cca 350
MtCO₂e in 2020. As shown in Figure 1.7, the biggest increment in the ICT
footprint is expected in 2020 by the mobile networks (domination of 51%).

Regarding telecoms devices (mobile phones, chargers, internet protocol
TV boxes and home broadband routers), global footprint was 18 MtCO₂e
in 2002 with an expected threefold increase to 51 MtCO₂e by 2020 (due to
growth in China and India) (Figure 1.8). With a 1.1 billion mobile accounts
in 2002, an expected increase to 4.8 billion in 2020 is the largest source of

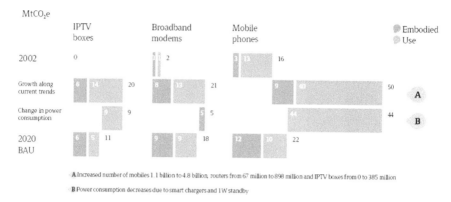

Figure 1.8 The global telecoms devices footprint.

global telecom footprint emissions. Increased access to broadband will also have an impact in the number of routers, which will grow from 67 million in 2002 to 898 million in 2020, as well as in the number of IPTV boxes (from none in 2002 to 385 million in 2020). Growth along current trends can be seen in row A of Figure 1.8. Unfortunately, the majority of emissions from mobile devices come from standby mode. This is related to so-called phantom power, and it is used by chargers that are plugged in but not in use. It is expected that "smart chargers" will be used for turning off when a device is not connected, as well as 1 W (or lower) standby standards. By following this assumption, a sharp decrease in charger consumption counteracts the growth in number of accounts, which brings an increase of the footprint of mobile phones to 4%. On the other hand, broadband routers and IPTV boxes increase their footprint comparatively more thanks to higher penetration from a small base today (Figure 1.8, row B).

Independently from the residential network, both wired and wireless, one can further think of telecommunications energy expenditure, especially due to the high mobile telecommunications global expansion. The power consumption in cellular networks infrastructure (base stations and core networks) doubles every year, and in 2007 it was approximately 60 TWh, with 80% of the energy consumption for the radio access network [40]. Thus, the footprint of the global telecoms infrastructure is shown in Figure 1.9. In 2002, it was 133 $MtCO_2e$, whereas in 2020 it is expected to 299 $MtCO_2e$ by 2020.

A key contributor to carbon emissions in 2020 will be mobile networks, driven largely by the increase in base stations and mobile switching centres. However, emissions from networks cannot be calculated based on the hard-

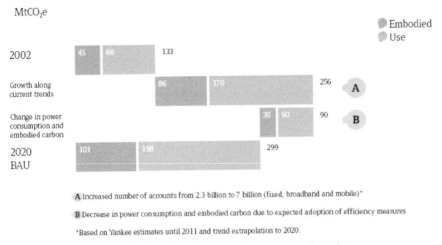

Figure 1.9 The global telecoms infrastructure footprint.

ware used in the network alone, nor are data available from each provider on specifically how much energy their networks consumed. For example, it is not currently cost-effective to implement night battery operation and solar-powered base stations can be used only in certain climates. Natural ventilation is already being used by some operators and would reduce the need to cool the base station and core network equipment. In addition, companies are experimenting with "network sharing", which reduces the need to construct new networks, and tracking the energy consumption reduction benefits.

A group of important energy conscientiousness directions aims at an increase of the energy efficiency of base stations and/or at a development of efficient energy harvesting models to run them and/or to apply combinations of currently available alternative naturally available energy sources and grid network. In this way, ICT applications can also contribute to a tangible impact on lowering greenhouse gas emissions, power consumption, and achieving efficient recycling of equipment waste, and thus promote environmental improvement [23].

1.4 ICT as Energy Saver and Enabler

It would be totally unfair to have an insight to ICT in the information era of the 21st century as "aggressor" in terms of large energy consumer for ICT devices and systems. Thus, the highly positive role of promoter and helper of energy saving has to be emphasized. There are already many initiatives around our globe for it. One important group of the intiatives is of FP7 European Commission, entitled "Towards Zero Power ICT" [41]. Its objective is efficient powering or self-powering of all the electronic devices surrounding us in daily life. The objective asks for a transformative research in sense of researching new energy-harvesting technologies, most probably at the nanometer scale, as well as their integration with low-power ICT into autonomous nano-scale devices for various purposes, e.g. sensing, processing, actuating and communications. Thus, self-powered autonomous nano-scale electronic devices that harvest energy from the environment would possibly combine multiple sources and store it locally. These systems would co-ordinate low-power sensing, processing, actuation, communication and energy provision into autonomous wireless nanosystems. In this way built autonomous nanoscale devices (from sensors to actuators) would extend the miniaturization of autonomous devices beyond the level of the "smart dust". These new systems would provide a vast number of innovative applications in e.g. intelligent distributed sensing, for health-safety-critical systems or environmental monitoring.

In addition to research activities, the Commission has recognized that ICTs and ICT-based innovations may provide one of the potentially most cost-effective means for Member States to achieve the 2020 target. The policy framework that will allow the energy-saving potential of wide-use of ICTs was initiated by adopting two Communications adopted in May 2008 [42] and in March 2009 [43]. The Recommendation [44] adopted by the Commission, on 9 October 2009, on "mobilising Information and Communications Technologies to facilitate the transition to an energy-efficient, low-carbon economy", aims to unlock energy efficiency potential through more public-private partnership initiatives, like the ones recently launched by the Commission on energy efficient buildings and green cars [45], but also through partnerships between the ICT industry and defined strategic sectors. In particular, the buildings, transport and logistics sectors are identified as key economic sectors where energy efficiency through the use of ICT is still largely untapped.

The global emissions reductions of 7.8 GtCO$_2$e in 2020, five times its own footprint have been identified in [37] (Figure 1.10).

In 2002, total emissions from human activity have been distributed as follows: 24% from the power sector, 23% from industry, 17% from agriculture and waste management, 14% from land use, 14% from transport and 8% from buildings. The consumption of electricity and fuel in 2005 has been

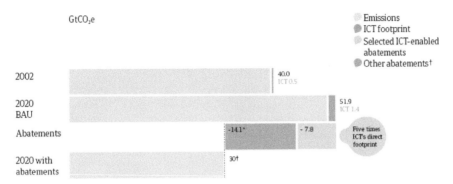

Figure 1.10 ICT impact: The global footprint and the enabling effect [38].

distributed to manufacturing (33% of end-use energy consumption), transport (26%), households (29%) and other services and construction (12%).

Figure 1.11 shows the enabling effect of ICT in all the mentioned areas. It has to be strenghtened that the savings are of 15% (as already mentioned, 7.8 GtCO$_2$e of the total 51.9 GtCO$_2$e).

In economic terms, the ICT-enabled energy efficiency translates into approximately EUR 600 billion of cost savings.

1.5 Conclusions

Emergency measures have to be taken for energy consumption for enabling further development of human society in protected global environment. The development of ICT at its current state-of-the-art supports high degree of transdisciplinarity with other engineering and non-engineering areas to achieve the "green environment" goal. Only in the very recent time, research for energy-efficient protocols and products has been initiated, but with the world being a global village, the excellent globally accepted and standardized solutions are to be expected very soon.

Predicting the pace and intensity of these virtualization trends is difficult, but the industry is well aware of the huge efficiency opportunity. Initiatives such as the "Green Grid", a global consortium dedicated to data centre efficiency and information service delivery, working towards new operating standards and best practices, has attracted support from the industry.

In its research agenda, ICT is already following bio-inspired algorithms, due to their evolutionary intelligence of promotion and development of the most efficient systems known today. The living organisms are efficient be-

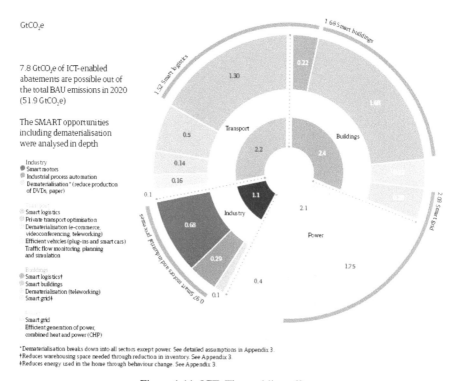

Figure 1.11 ICT: The enabling effect.

cause they are integrated, and are honed by evolution to be as economical as possible with the materials from which they are made and with substances like air, water and nutrients needed for life. This approach shall be mirrored also for energy consumption in all important areas, such as building areas, where building enclosure is noted and seen primarily as an energy barrier. By following bio-inspired energy paradigms together with newest ICT principles, such as context awareness [46], as the capability of perceiving the user situation in all its forms, and of adapting in consequence the system behaviour (i.e. the services and content supplied to the users), holistic concept of environmentally based organic architectures in the true, not stylistic, sense, will bring new values to our globe. This development of human race is possible only with the proper education of engineers in green engineering, with the primary goal to understand, perceive and consume holistic approach of surrounding environment, including properties of materials and living beings as well as the newest technological development. Only such an education

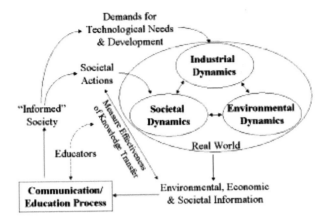

Figure 1.12 Role of education, knowledge transfer, and social processes in the creation of informed citizens and households [2].

could bring a new paradigm shift in engineering agenda toward proper "green engineering".

Thus, there is a great role of education, knowledge transfer and understanding of societal, industrial and environmental dynamics, as shown in Figure 1.12. Only by giving full respect to all three dynamics, the best results for the human race can be achieved.

The direct carbon footprint of the ICT sector is dominated by electricity consumption, so an obvious way to reduce emissions is to use as much electricity as possible from renewable sources. Thus, ICT companies can do this by applying several strategies: by purchasing renewable electricity, by installing renewable generation on their sites and by making renewable electricity integral to their products. Also, policy makers should be encouraged to create the right regulatory and fiscal environment to encourage investment in large-scale renewable generation. This will ultimately lead to a reduction in the "in use" phase of the ICT product life cycle. However, the natural position of ICT is to partner with power companies to optimise the existing electricity grid to allow more efficient power distribution and enable the use of more renewable or green power. There is a great possibility also for companies to use server space on demand to build their own applications and websites, the way one would pay monthly for electricity or water, known as "utility computing". These are both simple examples of what is more generally called "cloud computing", centralized and highly scalable services that could lead to further capacity to virtualise or consolidate resources with breakthrough gains in energy efficiency.

Except in its own sector, the dematerialization across public and private sectors could deliver a significant reduction of 500 $MtCO_2e$ in 2020, which is the equivalent of the total ICT footprint in 2002, or just under the emissions of the UK in 2007.

ICT provides an energetic road, which is not of physical, but of invisible-to-human-eye-nature, which covers the whole area with the speed of light. The ICT road does not connect only two points, i.e. it does not start at one and stops at the other point, but it gives possibility to everyone to be involved in the road at every moment the person wishes to be there.

The WPMC 2009, held in Sendai, Japan, is one of the first conferences which brought together four sessions on green communications, bringing together academic and industrial researchers for discussing energy-efficient commnications. It covered very broad range of topics from energy efficiency for IT to wireless sensor networks on the other side. The best ICT potential at this very moment is proving to be reduction of energy consumption in mobile and wireless access networks, broadband access networks and home/office/industrial ICT and residential networks, with the moving along holistic approach toward the "green world".

References

[1] World Commission on Environment and Development (WCED), *Our Common Future. The Brundtland Report*, Oxford University Press, Oxford, U.K., 1987.

[2] Mihelcic, J.R., Crittenden, J.C., Small, M.J., Shonnard, D.R., Hokanson, D.R., Zhang, Q., Chen, H., Sorby, S.A., James, V.U., Sutherland, J.W. and Schnoor, J.L., Sustainability science and engineering: Emergence of a new metadiscipline. *Environ. Sci. Technol.* **37**(23), 5314–5324, 2003.

[3] Lev-On, M., The road to sustainable development: From Rio to Johannesburg and beyond. *Environ. Manage., A&WMA*, May, 18–28, 2002.

[4] The Norwegian Nobel Committee, The Nobel Peace Prize 2007, Press Release, www.nobelprize.org/nobel_prizes/peace/laureates/2007/press.html, Oslo, 12 October 2007.

[5] Kyoto Protocol to the United Nations Framework Convention on Climate Change, United Nations, 1998.

[6] COM(2009)39, Communication from the Commission to the European Parliament, the Council, the European economic and social committee and the committee of the regions: Towards a comprehensive climate agreement in Copenhagen, SEC(2009) 101, SEC(2009) 102, Brussels, 28 January 2009.

[7] Directive 2003/87/EC, 25 October 2003.

[8] Directive 2004/101/EC, 27 October 2004.

[9] Directive 2008/101/EC, 19 November 2008.

[10] EC Regulation No 219/2009, 11 March 2009.

[11] Directive 2009/29/EC, 23 April 2009.

[12] Eckstein, A., Europolitics, Environment Council, 22 December 2009.

[13] Copenhagen Accord, FCCC/CP/2009/L.7, 15th Conference of the Parties, Copenhagen, 7–18 December 2009.

[14] Germany2020: *Future Perspectives for the German Economy*, McKinsey & Company, Inc., 2008.

[15] United States National Academy of Engineering, www.nae.edu.

[16] United States Environmental Protection Agency, www.epa.gov.

[17] Green Engineering: Defining the Principles, held in Sandestin, Florida, USA, May 2003.

[18] www.motorsmatter.org.

[19] www.energysmart.com.au.

[20] www.motor-challenge.eu.

[21] www.iec.ch.

[22] ICTs and Climate Change: ITU Background Report, ITU/MIC Japan Symposium on ICTs and Climate Change, Kyoto, 15–16 April 2008.

[23] ETNO and WWF: Saving the climate @ the speed of light, http://www.etno.be/Portals/34/ETNO%20Documents/Sustainability/Climate%20Change%20Road%20Map.pdf

[24] Intergovernmental Panel on Climate Change, IPCC graphs, WG1, Fig. SPM-2, www.ipcc.ch.

[25] Speech by the IPCC Chairman Mr R.K. Pachauri at the Welcoming Ceremony, IPCC at the United Nations Climate Change Conference-COP 15, Copenhagen, Denmark, 7 December 2009.

[26] US Department of Energy, Net Metering Policies, http://apps3.eere.energy.gov/greenpower/markets/netmetering.shtml.

[27] ITU Internet Report: The Internet of Things, ITU Strategy and Policy Unit (SPU), 2005.

[28] http://www.soumu.go.jp/joho_tsusin/eng/Releases/Telecommunications/news070406_1.html

[29] US Green Building Council, USGBC: Leadership in Energy and Environmental Design, LEED, 1998, www.usgbc.org.

[30] Sioshansi, F.P. and Davis, E.H., Information technology and efficient pricing-providing a competitive edge for electric utilies, *Energy Policy* 17(6), 599–607, 1989.

[31] Commission of the European Communities: Proposal for a Directive of the European Parliament and of the Council laying down the framework for the deployment of Intelligent Transport Systems in the field of road transport and for interfaces with other transport modes, COM(2008)887final, Brussels, 16 February 2008.

[32] European Commission, Information Society and Media, eCall-saving lives through in-vehicle communication technology, Factsheet, June 2008.

[33] Commission of the European Communities: Communication from the Commission to the the Council and Parliament: Green Paper on the Urban Environment, COM(90)218, Brussels, 27 June 1990.

[34] Priya, S. and Inman, D.J., *Energy Harvesting Technologies*, Springer, 2009.

[35] Stern, N., Stern review on the economics of climate change, Executive Summary, HM Treasury, Cambridge Press, London, 2006, retrieved January 5, 2010.

[36] Moore, G., Cramming more components onto integrated circuits, *Electronics Magazine*, 19 April 1965.

[37] The Climate Group: Global e-Sustainability Initiative (GeSI), SMART 2020: Enabling the low carbon economy in the information age, Creative Commons, 2008.

[38] Enkvist, P., Naucler, T. and Rosander, J., 2A cost curve for greenhouse gas reduction, *The McKinsey Quarterly* **1**, 2007.

[39] European Commission DG INFO, ICT for energy efficiency, Ad-Hoc Advisory Group Report, Brussels, 24 October 2008.

[40] Scheck, H.-O., Power consumption and energy efficiency of fixed and mobile telecom networks, ITU-T, Kyoto, April 2008.

[41] INFSO FET, R. Stuebner, Towards zero power ICT (2zeroP), National Contact Point Briefing, Brussels, 12 May 2009.

[42] Commission of the European Communities: Communication from the Commission to the European Parliament, the Council, the European Economic and Social Committee and the Committee of the Regions: Addressing the challenge of energy efficiency through Information and Communication Technologies, COM(2008)241, Brussels, 13 May 2008.

[43] Commission of the European Communities: Communication from the Commission to the European Parliament, the Council, the European Economic and Social Committee and the Committee of the Regions: On mobilising Information and Communication Technologies to facilitate the transition to an energy-efficient, low-carbon economy, COM(2008)111, Brussels, 12 March 2009.

[44] Commission of the European Communities: Commission Recommendation of 9.10.2009 on mobilising Information and Communication Technologies to facilitate the transition to an energy-efficient, low-carbon economy, C(2009) 7604, Brussels, 9 October 2009.

[45] Europa Press Releases RAPID, European Commission and industry to invest EUR 3.2 billion in economic recovery for a stronger, greener and more competitive economy tomorrow, IP/09/1116, Brussels, 13 July 2009.

[46] Chaari, T., Dejene, E., Laforest, F. and Scuturici, V.-M., Modeling and using context in adapting applications to pervasive environments. In *ICPS'06: IEEE International Conference on Pervasive Services*, 2006.

2

How Telecom Could Save the Planet

L.E. Mägi, M.G. Gustafsson and A. Kramers

Ericsson, Stockholm, Sweden

Abstract

In recent years, urbanization has increased rapidly creating an ecological imbalance on planet earth. There is a growing need to design sustainable cities dedicated to minimize its required inputs and waste outputs. ICT's new system solutions provide scope for synergies between waste and energy production and enable coordination with efficient land use, landscape planning and transport systems.

Keywords: Sustainable city, Intelligent Transport Systems (ITS), telemedicine, e-learning.

2.1 Introduction

What is a sustainable city? Why do we need to start building them now? City planners have their work cut out for them, because urbanization is accelerating. The year 2007 marked the first time in history that more than half the world's population lived in cities. Mega-cities mean mega-problems for our survival.

2.2 The Challenge

India has 10 of the 30 fastest-growing urban areas in the world and, based on current trends, analysts estimate that 700 million people – roughly equivalent to the current population of Europe – will move to cities in India by 2050.

R. Prasad et al. (Eds.), Towards Green ICT, 29–36.

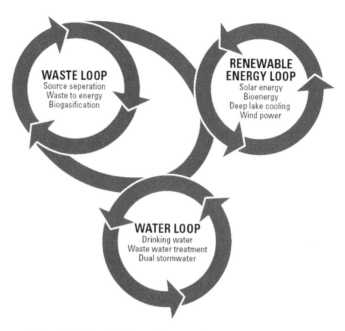

Figure 2.1 The shift form linear to circular resource flows.

In China, 400 million people will move to cities over the next 20 years. These huge demographic changes, combined with the increasing ecological pressure that humans put on our planet, will spur the need for cities that carry their own weight.

Information and communication technology (ICT) has an important role to play in reducing CO_2 emissions. Globally, the ICT industry itself emits about 2% of total emissions but has a long-term potential (approximately 2050) to reduce those emissions by 7–35%. In the short to medium term, a reasonable reduction level for ICT is estimated at 5–20% [1].

A sustainable city is an entire city dedicated to minimizing its required inputs (energy, water, and food) and waste outputs (heat, air pollution, methane, and water pollution). A sustainable city can feed and power itself with minimal reliance on the surrounding countryside, and creates the smallest possible ecological footprint for its residents (Figure 2.1). Such a city produces the least pollution, attains highly efficient land use, and makes a minimal contribution to global warming.

Figure 2.2 The shift form linear to circular resource flows.

Cities are best placed to lead in reducing worldwide emissions. The way we plan, design, and manage cities provides solutions to adapting to a changing climate and mitigating any further changes.

Each cluster in the city should strive to have complete cycles for its own water, cleaning, food, and energy. It will become essential to use building façades and roofs for food production and water collection.

In a sustainable city, use of renewable resources is emphasized, consumption is minimized, and resources are managed in a way that maximizes recovery and reuse. New system solutions provide scope for synergies between waste and energy production and enable coordination with efficient land use, landscape planning, and transport systems.

Swedish architectural firm Tengbom has won an international contest for an expansion of the City in Chonhqing in China. The area is more than 3 million km^2, includes residential, offices and commercial buildings and is part of urban planning project for the whole of Chongqing. The aim is to preserve the natural conditions, topography and vegetation (Figure 2.2).

2.3 ICT: Friend or Foe?

ICT emissions are projected to double by 2020, to approximately 4% of global CO_2 emissions. At the same time, ICT can enhance existing processes to make products better and make people and machinery more efficient.

ICT involves monitoring and analytic tools, logistics, intelligent transport systems including tracking and telematics, smart building technologies, and energy management systems and technologies that manage user behavior and intervene intelligently to minimize energy waste.

ICT can enable new ways of working by providing energy-related applications that facilitate renewable power generation and virtualization or dematerialization of goods and services, such as paperless office technologies. It can also contribute by transforming the way we interact – virtual conferencing, for example – including all the tools that help us make better decisions.

Urban planning represents the interaction between the different parts of society and the opportunity to make things flexible so they can constantly be redesigned. It will enter a new era where little central planning is required yet still the system can interact and develop in a more resource-efficient direction.

Values and attitudes are often ignored when sustainability is discussed. But at the end of the day, they are what will guide us and determine what a future society will look like. In order for ICT to be used to its full potential, cities must include ICT infrastructure as they develop. For example, better broadband and connectivity availability will enable teleworking and e-learning via video conferencing.

The matter of social identity is also important to city planners. To attract citizens, a skilled workforce, and investors, city planners need to understand and consider how people's values are reflected in design, functions, communication concepts, and type of city.

2.4 New and Existing Urban Ideas

The sustainable city can be divided into five spheres of influence: city, block, house, household, and citizen. We will give examples connected to the spheres of ICT solutions that can have a big impact on greenhouse gas emissions. Road transport represents a substantial source of all emissions. When it comes to CO_2 in Japan, for instance, the sector is responsible for almost 20%, and in Sweden for 30%. Intelligent transport systems (ITS) promote a more effective logistic flow, and a better functioning and more used public transportation system.

ITS can be described as a network where people, roads, and vehicles are linked, exchanging information among cars, drivers, pedestrians, and other users.

Japan was an early implementer of ITS [2]. The Japanese government has promoted cooperation between different organizations and initiated collaboration projects such as "free mobility assistance", launched by the Ministry of Land, Infrastructure, Transports and Tourism. The project aims to create a network society where information terminals can communicate road descriptions, guides, and traffic information. This information can be sent to a mobile phone and received as a voice message in different languages.

Intelligent transport solutions could have negative, so called rebound, effects. The aim is to reduce emissions and not to make it more efficient to use transportation and hence increase the amount of transportation.

Utilities are now turning their attention to initiatives that integrate the electric power grid and back-office processes. As an example, Italy's second largest electricity distributor Acea recently outsourced the management of its automatic meter reading to Ericsson. The system is being deployed in Rome to 1.5 million households and allows Acea's customers to get invoices based on actual consumption levels and use different tariffs. Households are linked via the power lines to concentrators, which in turn are linked via radio modules to the service center where data are integrated with Acea's systems for billing, network management and customer care. Fever and shorter power breaks are expected due to immediate notification. This solution lowers costs for utilities in customer management and network maintenance, as well as protecting revenues by more accurate billing and early detection of "power fraud."

The utility of the future, in which intelligent grids and intelligent enterprises are tightly integrated, will require wireless, wireline, and other forms of communication. Leveraging telecommunication architectures and expenditures is of paramount importance if utilities are going to succeed in implementing advanced metering infrastructure, demand response, and distributed generation from hundreds of small power-generating units instead of a few large ones, as well as other applications requiring advanced intelligence in the smart grid.

2.5 Taking Care at Home

Various mobile telemedicine systems can be used to deliver healthcare to citizens at home, saving the patient a trip to the hospital; for instance, remote meetings via video conference, sensors connected to mobile phones measuring vital body parameters such as respiratory rate, heart rate and blood pressure, as well as homecare equipment for blood analysis.

One example of how mobile telemedicine can be used to reduce CO_2 emissions is homecare for patients with leg and foot wounds. In a case study conducted by Ericsson Research, the number of hospital consultations was estimated to be reduced by 50% by introducing telemedicine. The emissions from the telemedicine system were compared to the reduced emissions due to less travel. The resulting yearly CO_2 net reduction was 63 kg for an average Swedish homecare patient with leg wounds. The introduction of the telemedicine system increased CO_2 emissions marginally but by reducing the need for travel, it saved 20 times that amount.

If mobile telemedicine were introduced nationwide, for example in Sweden for the entire patient group the total CO_2 reduction would be 2100 metric tons yearly, with a total transport cost saving of about USD 11.5 million. These figures are only for one of many potential patient groups and care processes where mobile telemedicine can be introduced to reduce CO_2 emissions. For a large emerging city the potential to leapfrog conventional care processes to reduce environmental impact is enormous.

But it is also a viable way to provide medicine to rural, remote areas. In India with 72% of the population living in rural areas, but 80% of the doctors living in urban areas, the relevancy of a service that enables remote access to healthcare is clear. When it comes to secondary and tertiary care, 90% of the facilities are in the towns and cities far away from the rural India, making the need for new alternatives even greater [3]. As the outreach of the primary health centers is inadequate [4], the level-m telemedicine can be used to close the largest gaps in the current infrastructure. The Indian government's 11th five-year-plan (for 2007–2012) INR 2000 million was allocated to telemedicine initiatives. Most of this funding will be channeled to public-private partnerships [5].

2.6 Smarter and More Efficient

Flexi-work and e-learning require collaboration tools. Phone meetings are commonly used but video will be used more and more to support flexi-work and e-learning.

For videoconferencing to be a trusted application, it must be easy to use and reliable. The most important aspect is audio quality, but good picture quality allows participants to use body language in a way that contributes value to the conference. To secure this, higher connection speeds are often needed.

E-learning can be performed in different ways: "live," webcast, or archived sessions, depending on whether the audience needs to interact with the teacher directly, or webcast sessions where e-mail contact or chat is possible. In a block within a sustainable city, other applications can support a more efficient use of products, for example, cars and washing machines. Citizens can book cars in a pool system, coordinate laundry time, or reserve other shared facilities through their mobile devices.

Intelligent buildings are designed for long-term sustainability and minimal environmental impact through the selection of recycled and recyclable materials, construction, maintenance, and operations procedures. The ability to integrate building controls, and optimize operations and enterprise-level management results in a significant enhancement in energy efficiency, lowering both cost and energy usage. ICT can monitor water and energy consumption, lighting, facilities, security systems, parking, access, and surveillance cameras.

In the household, automation can include the control of electronic equipment, water usage, heating and cooling, as well as multimedia home entertainment systems such as video-on-demand. Other examples are automatic plant watering, pool pumps and scenery for dinners and parties. In all of these scenarios the mobile phone would serve as the remote control.

We also want to make informed decisions in our role as consumers and decide for ourselves when and how to travel, whether to buy locally produced products and services, and share rather than own everything we use. To support us in making the best decisions, technologies such as near field communication (NFC) and radio frequency identification (RFID) provide solutions. Holding a sensor such as a mobile phone close to a product would generate information about the product including method of transport, country of origin, factory or farm of origin, whether the product is ecological or not, and a list of ingredients. Continuous internet access is another information channel to be used by citizens wanting to update themselves on the latest sustainability aspects.

2.7 A Holistic Systems Approach

Taking on a project as complex as a sustainable city requires finding the right working model to achieve close cooperation between different stakeholders. Support from government on all levels to stimulate cooperation through incentives, standards, regulation, and defining clear goals is necessary. Of course, all cities are unique and therefore need many customized solutions;

and an existing city that wants to become sustainable has different challenges than a completely new city.

On a practical level, some of the complex ICT systems solutions described above need new organizational thinking. Outsourcing telecom equipment and the running of networks to operators and service providers is a way to increase specialization, raise competence levels, and gain benefits of scale. However, the intention of ICT solutions is to give results; just buying the latest technology without a clear vision of a sustainable city will not make the city sustainable.

To understand what solutions to choose, it is essential to stay up-to-date with the latest information and use system tools such as Life Cycle Analysis calculations to include a complete system view and avoid sub-optimization and short-term decisions.

Environmental goals often support other goals. When reducing transport, for instance, one of the major advantages besides environmental benefits is cost reduction. For telemedicine systems, reduced transports can most likely finance both the telemedicine system and its operation and still result in a net saving and a better life for the patients.

The future of our cities and their citizens depends upon the decisions made by us today!

References

[1] Can telecom save the planet? *Ericsson Business Review* **2**, 2008.
[2] ITPS R2008:006 IT and the environment – Current initiatives in Japan and USA, Report, 2008
[3] Ministry of Health & Family Welfare, National Rural Telemedicine Network, Suggested Architecture and Guidelines, Draft Proposal, Ministry of Health & Family Welfare, 2008, www.mohfw.nic.in.
[4] Sood, S., Mbarika, V, Jugoo, S., Dookhy, R., Doarn, C.R., Prakash, N. and Merrell, R.C., What is telemedicine? A collection of 104 peer-reviewed perspectives and theoretical underpinnings. *Telemedicine and e-Health* **13**(5), 2007.
[5] Solberg, K.E., Telemedicine set to grow in India over the next 5 years. *The Lancet* **371**, 2008.

3

Directions towards Future Green Internet

Hideaki Imaizumi and Hiroyuki Morikawa

Research Center for Advanced Science and Technology, The University of Tokyo, Tokyo 153-8904, Japan

Abstract

This paper presents several technical directions towards realizing future green Internet. The threat of man-made climate change has become a key issue in the world and is currently forcing various industries including ICT to reduce both energy consumption and carbon emissions. This paper describes power consumption and carbon emissions caused by ICT and their forecast in the future. In addition, we will clarify which parts of ICT, especially Internet, consume the majority of power and will discuss various challenges about the future green Internet. Finally, we will introduce our research activities for reducing power consumption towards realizing the future green Internet.

Keywords: Green Internet, energy-efficient networking.

3.1 Introduction

Global warming, the threat of man-made climate change, has become a key issue in the world and is currently forcing various industries to reduce both energy consumption and carbon output; the Information and Communication Technology (ICT) industry is no exception due to the fact that ICT has a significant and growing impact on power consumption and carbon emissions.

According to a report by the Climate Group, the total carbon emissions related to ICT in the year 2007 are approximately 2% of the worldwide carbon emissions [1]. Telecom infrastructures and devices cause 37% of the total

R. Prasad et al. (Eds.), Towards Green ICT, 37–53.

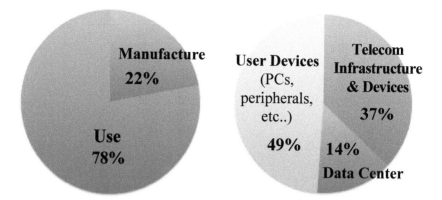

Figure 3.1 Carbon emissions in ICT.

carbon emissions in ICT, while the remains are caused by data centers and user terminals as depicted in Figure 3.1 [1]. The ratio of the emissions in the network infrastructures to the total ICT emissions will significantly increase according to a number of projections that Internet traffic will enormously increase due to the emergence of bandwidth-intensive applications such as 8K Super Hi-Vision by NHK [2].

According to the Japanese Ministry of Economy, Trade and Industry (METI) [3], the average Internet traffic volume observed in Japan for the year 2006 was 637 Gbps and projections indicate that by 2025, it will exceed 121 Tbps; projections also indicate that network-related power consumption will increase by approximately 13 times in the same time span. This will cause a serious impact on power consumption and carbon emissions. The National Institute of Advanced Industrial Science and Technology (AIST) in Japan estimates that the ratio of the total power consumed by only routers out of the total electricity generated in 2005 was less than 1.0% in Japan. If the amount of total electricity generation remains, the ratio will grow up to 1.7% by 2010, 9% by 2015 and 48.7% by 2020, respectively [4].

The traditional method for reducing power consumption is to manufacture chips using smaller semiconductor fabrication technologies such as those used in the recent move to 32 nm process shown in the Intel processors. However, it is well known that increasing current leakage prevents this technique from being carried out much further [5]. Therefore, much more research should be conducted towards reducing power consumption in Internet routers.

In this chapter, we discuss possible means for reducing power consumption caused by the usage of Internet and try to give several directions towards realizing a future green Internet. First, we clarify which parts of the network components mainly consume the majority of power in network infrastructures in Section 3.2. Based on the discussion, we introduce four directions for efficiently reducing power consumption in Internet and introduce existing challenges for each direction in Section 3.3. Then, we introduce our research activities for this issue in Section 3.4 and summarize this chapter in Section 3.5.

3.2 Power Consumption in Internet

In this section, we explore power consumption caused by Internet and introduce two indexes for evaluating energy efficiency in Internet. Finally, we introduce the breakdown of power consumed by routers as one of the key devices in Internet.

Excluding leaf networks such as Home/Office networks, Internet consists of a huge number of Internet Service Providers (ISPs) inter-connected in Internet eXchange points (IXes). Usually, a whole network in an ISP and an IX is divided into multiple local sub-networks and each of them is located in different areas in order to provide services widely. Each local sub-network consisting of ICT devices such as routers, switches and servers is usually operated within a rented space in a building. In this chapter, we call such spaces Network Bases (NBs).

It is very important to consider power consumption caused by NBs due to the fact that overhead factors such as cooling, power conversion, and lighting consume significant power. An example of the impact caused by such overhead factors in data centers in the US is illustrated in the left-hand side of Figure 3.2. According to this figure, 63% of the total power is consumed by the overhead factors. Although NBs consist of different devices, the situation in NBs would be not so far from this case.

In data centers, an index called Power Usage Effectiveness (PUE) is used to evaluate the impact on power consumption caused by the overhead factors. It is calculated as

$$\text{PUE} = \frac{\text{Total Power Consumption}}{\text{Power Consumption of ICT}} > 1.0 \qquad (3.1)$$

PUE in data centers is normally very poor as illustrated in the left-hand side of Figure 3.2. The average PUE in data centers in Japan is between 2.2 and 2.5 [7]. A remarkable 1.12 PUE was observed in Google data center F as depicted in the right-hand side of Figure 3.2 [8]. This result has been achieved

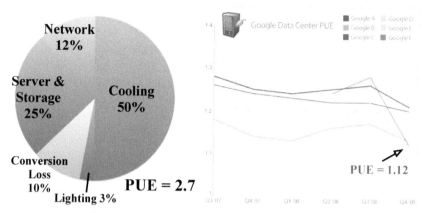

Figure 3.2 Power usage effectiveness in data centers.

with water radiators connected to a river and their improvement on power supply.

However, in the case of networks, it is difficult for NBs to be flexibly located close to rivers similar to the data centers. Therefore, we need to pay attention to PUE in NBs as well in order to fairly evaluate the total power consumption in NBs.

On the other hand, power consumption caused by ICT devices in NBs is also important to be explored. According to the report from METI, the majority of power consumption in the Internet will be consumed by routers [3]. Therefore, we concentrate on the power consumption caused by routers.

In order to evaluate energy efficiency in IP routers, an index called Energy Efficiency Rate (EER) is used to represent the delivered full-duplex throughput (bps) by the number of watts (W) required to produce that throughput and is calculated as [6]

$$\text{EER} = \frac{\text{Full Duplex System's Capacity (bps)}}{\text{System's Total Power Consumption (W)}} \qquad (3.2)$$

Table 3.1 shows EER performances of several routers [5, 6]. EERs of recent routers with improvement on energy efficiency such as reduction of the number of chips with centralized chip design are better than Cisco CRS-1 [5]. Cisco CRS-1 was released in 2004 while Alaxala AX6308S and Juniper T1600 were released in 2006 and 2007, respectively.

In order to find effective ways to reduce power consumption in routers, we need to clarify which components consume much power. Table 3.2 shows breakdown of power consumed by a router [9]. Single-chassis and Multi-

Table 3.1 Energy efficiency rate examples.

Router	EER
Cisco CRS-1 (2004)	50 Mbps/W
Juniper T1600 (2007)	163 Mbps/W
Alaxala AX6308S (2006)	175.4 Mbps/W

Table 3.2 Breakdown of power consumed by a router [9].

	Percentage of Total Power	
	Single chassis	Multi-chassis
Supply loss and blowers	35%	33%
Forwarding engine	33.5%	32%
Switch fabric	10%	14.5%
Control plane	11%	10.5%
I/O (O/E/O)	7%	6.5%
Buffers	3.5%	3.5%
Total	100%	100%

chassis routers indicate Cisco 12816 edge routers and Cisco CRS-1 carrier routing systems, respectively.

Out of the total power, 33–35% is consumed by miscellaneous components such as power supply inefficiency, fans and blowers. The majority of the power consumption is consumed by four functional elements in data plane: forwarding engine, switch Fabric, I/O including O/E/O conversion, and buffers consume 33, 10–15, 7, and 3.5% of the total power consumption, respectively, while the power consumption in control plane is approximately 11%. In disregard of the miscellaneous components, forwarding engine is the functional element consuming the majority of the total power in routers.

3.3 Challenges for Green Internet

Considering a future Internet will contain a huge number of user devices such as mobile phones, sensors, femto cells, access points and TVs, we need to pay attention not only to routers but also to user devices. We herein discuss effective directions for reducing the power consumption in the Internet.

In order to realize the future green Internet, we believe there are four directions: (1) Profiling Platform, (2) Activity-adaptive Internet Architecture, (3) Low Power Forwarding Mechanism, and (4) Energy Harvest. Each of them is complementary and the concurrent progress will lead synergy effects on further reduction in the total power consumption. We will discuss each of them in the following sections.

3.3.1 Profiling Platform

In order to evaluate performance of various approaches for reducing the power consumption, a profiling platform which measures power consumption in target networks will be a key element.

The total energy efficiency in an NB can be represented by the following indexes inherited from the indexes described in the previous section:

$$\text{NB PUE} = \frac{\text{Total Power Consumption}}{\text{Total Power Consumption of Network}} \quad (3.3)$$

$$\text{NB EER} = \frac{\text{Total Network Throughput (bps)}}{\text{Total Power Consumption (W)}} \quad (3.4)$$

NB PUE is mostly same as PUE although it limits to network devices. On the other hand, NB EER expands EER to a network. The total network throughput of NB EER can be the amount of current input/output traffic measured at all external links. These indexes will give us significant overview of the energy efficiency in the NB in terms of not only its power usage effectiveness but also its relationship between network performance and power consumption.

However, the profiling platform should measure not only such total energy efficiency but also power consumed by each component in the devices and clarify the relationship between power consumed by each component and traffic load at each moment. This will allow us to analyse which parts of the target network really consume the majority of the power consumption in real-time and give us the next direction for further improvement.

The real-time information from the platform would potentially lead novel technologies such as energy-aware routing algorithms and dynamic network reconfiguration. Moreover, the relationship between such information and upper-layer services would give us significant influence. If we can obtain power consumed by a mail or even a packet and show users the result, it would encourage users to migrate to new green-aware protocols and software, and would lead to new kinds of services.

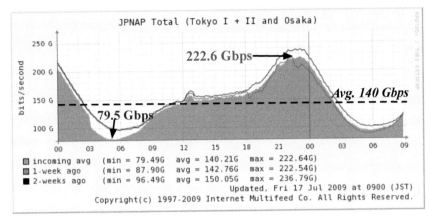

Figure 3.3 Traffic load during a day at IX.

3.3.2 Activity-Adaptive Internet Architecture

According to a report, average Internet backbone and LAN utilization is approximately 15 and 1%, respectively [10]. In addition, as Figure 3.3 illustrates, the total amount of traffic volumes observed at an IX in Japan indicates that traffic volumes vary widely even in IXes, depending on time, namely human activities during a day.

Therefore, by forcing performance of various elements consisting of the Internet to adapt to such activities, the total power consumption in the Internet can be reduced. In order to efficiently reduce the power consumption, an activity-adaptive Internet architecture where Internet devices such as routers, switches, optical network units (ONUs), femto cells, access points (APs), and various user devices cooperate with each other will be necessary.

The activity-adaptive Internet architecture can be divided into three levels: link/node level, network level, and upper-layer protocol level.

The link/node level in the architecture provides fundamental functions for performance adapting, sleeping, and resuming in links and nodes in accordance with traffic rate. In Internet backbone, dynamic performance adaptation such as link rate, internal clock frequency in routers/switches to the activities is suitable due to the fact that traffic always flows in backbones. On the other hand, in Home/Office environments, traffic could be intermittent because the number of people using the network is limited. Therefore, links and nodes such as broadband routers, switches, ONUs, and APs can transition from active into sleep-mode. In order to effectively reduce power consumption without any degradation of quality, efficient algorithms for estimating activity change and rapid wake-up mechanisms will be necessary. Sensor network technologies will be useful in order to infer human activities in Home/Office environments.

The network level provides dynamic topology reconfiguration using the functions provided in the link/node level. In the case that traffic rate is decreasing, an activity-adaptive routing protocol shrinks the topology by forcing traffic flows to be aggregated into particular paths and making nodes not relative to the paths to transition into sleep-mode. Otherwise, the routing protocol expands the topology by forcing necessary nodes in sleep-mode to transition into active-mode and distributing traffic into newly available paths. The routing protocol must be carefully designed with deep consideration to several parameters such as wakeup delay.

One of the major issues in the architecture is the impact to upper-layer protocols which define hello or keep-alive messages for confirming the connectivity between two nodes. Even if a user node and a corresponding local network are in sleep-mode, hello or keep-alive messages transmitted from another network will trigger transition of nodes in the network into active-mode. The activity-adaptive architecture would force such protocols to be redesigned or require other solutions such as the use of proxies.

A variety of activity-adaptive approaches have been proposed. Gupta's seminal work [11] in this area proposed an activity-adaptive approaches for links, switches, and routers, and demonstrated the existence of inter-packet gap for reducing power consumption. Later approaches include dynamic Ethernet link or switch shutdown during non-existence of incoming traffic [12–14], dynamic adaptive link rate for incoming traffic [16,17], an approach combining these two approaches [18], dynamic adaptation of internal clocks in switches [5] and shrinkable/expandable virtual networks with live router migration [19].

3.3.3 Low Power Forwarding Mechanism

The majorities of the power consumption in routers are caused by the forwarding engine and switching fabric as described in Section 3.2. In order to reduce the power consumption in routers, it is very important to develop a lower power forwarding mechanism.

One of the reasons why the IP forwarding mechanism consumes a large portion of the power consumption is that it processes incoming packets in a per-packet manner. Therefore, the increasing number of packets caused by the evolution of link bandwidth leads to much more power consumption. Especially, Ternary Content Addressable Memory (TCAM) used for searching a next-hop address for each incoming packet is one of the devices consuming a large amount of power due to its concurrent searching mechanism. The number of TCAMs used in routers will increase due to its increasing necessity for flow identification for providing various transport services such as QoS-guaranteed transport in the future Internet.

One of the major approaches is to remove any per-packet processing from routers by introducing either TDM or circuit switching into some parts of the networks. A flow is bound to a particular time slot or circuit with a signaling mechanism such as Generalized Multi-Protocol Label Switching (GMPLS) and any intermediate router within the network forwards packets in a time slot or circuit basis. Due to its affinity to circuit switching, optical circuit switching (OCS) has been researched as one of the promising technologies and its power consumption can be almost four orders of magnitude lower than that in IP routers [20]. Several other approaches exploit applying pseudo-TDM into IP networks [21, 22].

Another approach is removing O/E/O conversions and bit-rate dependent processing in routers by introducing optical switching into the packet switching architecture. While semiconductors used in current routers consume much higher power at higher bit-rate, the power consumption of optical devices do not depend on it. As an example, Multi-Wavelength Optical Packet Switching (MW-OPS) is expected to reduce power consumption particularly caused by I/O and switch fabric in IP routers due to the property that it is capable of reducing the number of optical devices by switching a wavelength-multiplexed optical packet with a single wideband optical switch [24, 25].

Several researches are trying to reduce power consumption in routers without any modification to the forwarding mechanism by introducing super-conductivity based on SFQ (Single Flux Quantum) devices [26, 27]. SFQ devices can operate at over 100 GHz with 10^{-7} W while current semiconductor devices driven at high clock frequency over 10 GHz cause difficulty in integration due to its heat density and extremely high power consumption.

3.3.4 Energy Harvest

One of ecological ways for operating ICT devices without any carbon emissions is to harvest energy such as sonic pressure, foot pressure, vibration, solar, and wind.

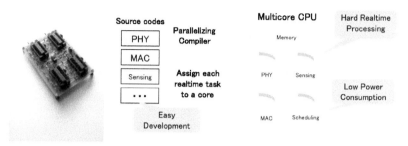

Figure 3.4 Multi-core CPU for WSNs.

In the future Internet, various kinds of devices such as mobile devices, sensors, access points, and femto cells will connect to the Internet. Harvesting energy in the ecological ways can contribute longer battery life. As an example, the ULP project in Japan is currently developing full wireless terminals without battery and the terminals that work with energy produced from vibration [28].

Moreover, it is potentially possible for large-scale networks to harvest such energy. In fact, NTT, one of the major carrier companies in Japan, is trying to harvest energy from solar power and wind to reduce the total power consumption in NTT [29].

3.4 Approaches

In order to reduce power consumption in Internet, we have conducted various researches. Here, we introduce our approaches in the following sections.

3.4.1 Multi-Core CPU for Wireless Sensor Networks

In order to monitor human activities in Home/Office environments and control user network devices such as broadband routers, WiFi access points and ONU devices, wireless sensor networks are very useful for detecting user contexts.

Wireless sensor nodes (WSNs) are usually designed based on single-core CPU architecture. However, they consume much power when complex concurrent tasks requiring higher clock frequency run on such nodes due to the fact that the power consumption is roughly proportional to the clock frequency cubed.

Figure 3.5 Solar biscuit prototypes.

In order to reduce the power consumption in WSNs, we apply multi-core CPU architecture into WSNs as illustrated in Figure 3.4 [30]. The results show that a sensor node with triple CPUs can eliminate about 76% of the power consumption compared to a single CPU sensor node. Moreover, this enables users to easily manage hard real-time tasks in a multi-core programming manner.

3.4.2 Solar Biscuit

In order to reduce the power consumption in WSNs, we have developed a battery-less wireless sensor node called Solar Biscuit which harvests energy using a solar panel to maintain semi-permanent availability as depicted in Figure 3.5 [31]. Our challenge is to design an appropriate system (communication mechanism, task scheduling, etc.) adapting to more unstable power source than batteries. We are currently designing a communication protocol suitable to utilize unstable harvesting energy

3.4.3 Ultra Low Power Wakeup for Wireless Communications

Excessive power consumption is a major problem in wireless communication, since wireless devices consume a considerable amount of energy in idle listening. Wake-up wireless communication technology is a promising candidate for reducing power consumption during idle listening.

To realize wake-up wireless communication, we develop a novel wake-up mechanism based on identifier matching. Furthermore, we consider applying a Bloom filter to the identifier matching as shown in Figure 3.6 [32]. We design and implement a wireless wake-up module that uses this ID matching mechanism. Simulation results reveal that the wake-up module consumes

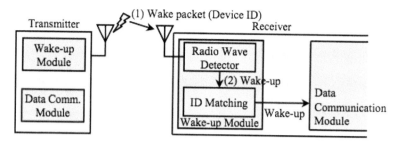

Figure 3.6 Two-step wakeup wireless communication.

only 12.4 W while it is in idle listening mode, and that employing this Bloom-filter-based approach eliminates 99.95% of power consumption in our application scenarios.

3.4.4 Power-Saving Technique Based on Simple Moving Average for Multi-Channel Aggregated Links

Several datalink technologies such as IEEE 802.3ad Link Aggregation and IEEE 802.3ba 40/100 GbE exploit multiple data channels for a single logical aggregated link.

In order to reduce power consumption of the aggregated links, we apply an activity adaptive approach to the aggregated links [34, 35]. Our approach changes the number of active channels belonging to an aggregated link based on a simple moving average in accordance with the current rate of traffic outbound onto the aggregated link. We have proposed an algorithm for estimating an appropriate number of active links based on the traffic rate and evaluated its performance. The results show that the algorithm can reduce the average number of active channels by a maximum of 40–55% without sacrificing buffering delay. Currently, we are improving the algorithm and trying to achieve an optimized solution.

3.4.5 Multi-Wavelength Optical Packet Switching

The recent progress of optical transport technologies especially on WDM and multi-level modulation technologies enables a huge capacity of 32 Tbps within a single fiber and this will cause unrealistic power consumption in routers [33].

In order to address this issue, we have proposed a Multi-Wavelength Optical Packet Switching (MW-OPS) architecture. As described in Section 3.3,

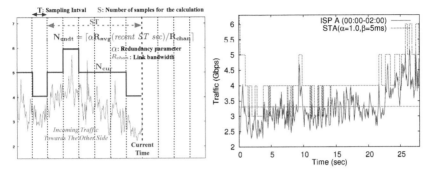

Figure 3.7 The algorithm model and the result.

MW-OPS is to reduce power consumption particularly caused by switch fabric in IP routers due to the property that it is capable of reducing the number of optical devices by switching a wavelength-multiplexed optical packet with a single wideband optical switch.

We are developing and implementing our MW-OPS switching nodes with contention resolution mechanisms using various optical switches such as PLZT optical switches, SOA switches, and current-injection total-reflection optical switches. We are also evaluating its performance in terms of bit error rate, throughput, and power consumption. Recently, we achieved 320 ($32\lambda \times 10$) Gbps MW-OPS switching with contention resolution mechanisms based on fiber delay lines as illustrated in Figure 3.8 [25].

3.4.6 Hybrid Optical Network Architecture

The future Internet will require not only high capacity but also QoS-guaranteed transport, high bandwidth utilization, multicasting and low power consumption.

In order to satisfy such requirements for the future Internet, a network architecture based on Optical Circuit Switching (OCS) would not be sufficient due to the nature of circuit switching that it provides an optical circuit for each packet flow and it causes significant waste of bandwidth when accommodating a huge number of small packet flows caused by interactive communication.

To address this issue, we design a novel Hybrid Optical Network Architecture (HOTARU), which combines both OCS and MW-OPS. In a network based on the architecture, OCS provides QoS-guaranteed communication and MW-OPS provides interactive communication. We are investigating flexible

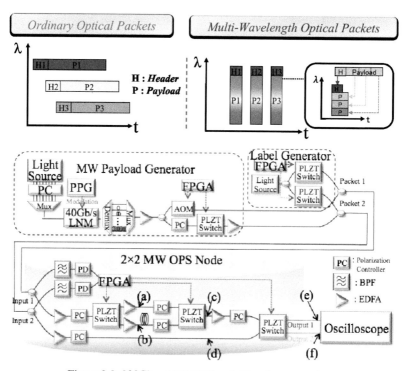

Figure 3.8 320Gbps MW-OPS switching demonstration.

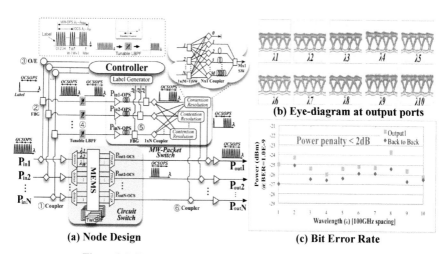

Figure 3.9 Demonstration of hybrid optical switching node.

wavelength assignment algorithms, routing algorithms, differentiated service provisioning mechanism, and the design of core node, and evaluating the performance via our simulator. We recently implemented 400 ($10\lambda \times 40$) Gbps hybrid optical switching node with dynamic resource allocation as illustrated in Figure 3.9 and we confirmed its feasibility [23].

3.5 Summary

In order to reduce power consumption and carbon emissions in Internet, this chapter presented several technical directions towards future green Internet. First, this chapter described power consumption and carbon emissions caused by ICT and their forecast in the future. In addition, we clarified which parts of ICT, especially Internet, consume the majority of power and discussed directions for improvement towards future green Internet. Finally, we introduced our research activities addressing this issue.

References

[1] SMART 2020 Report. Enabling the low carbon economy in the information age, `http://www.theclimategroup.org/assets/resources/publications/Smart2020Report.pdf`.

[2] Asami, T., et al., Energy consumption targets for network systems, Tu.4.A.3, ECOC2008.

[3] Hoshino, T., Plenary talk, Green IT Symposium, August 2007.

[4] Hazama, T., Symposium on Information and Energy, March 2006 Feature Article 01, `http://www.nistep.go.jp`.

[5] Yamada, M., et al., Technologies to save power for carrier class routers and switches, SAINT 2008, July 2008.

[6] Ceuopens, L., et al., Power saving strategies and technologies in network equipment opportunities and challenges, risk and rewards, SAINT 2008, July 2008.

[7] Suwa, K., DB1-4, Next generation data center 2009, July 2009.

[8] Google, `http://www.google.com/corporate/green/datacenters/measuring.html`.

[9] Baliga, J., et al., Photonic switching and the energy bottleneck. In *Proc. IEEE Photonics in Switching*, August 2007.

[10] Odlyzko, A., Data networks are lightly utilized, and will stay that way, *Review of Network Economics* **2**(3), 210–237, September 2003.

[11] Gupta, M., Greening of the Internet, ACM SIGCOMM 2003, August 2003.

[12] Gupta, M., Using low-power modes for energy conservation in ethernet LANs, IEEE INFOCOM 2007, May 2007.

[13] Gupta, M., Dynamic ethernet link shutdown for energy conservation on ethernet links, IEEE ICC 2007, June 2007.

[14] Gupta, M., A feasibility study for power management in LAN switches, IEEE ICNP 2004, October 2004.

[15] Tamura, H., et al., Performance analysis of energy saving scheme with extra active period for LAN switches, IEEE Globecom 2007, November 2007.

[16] Gunaratne, C., Reducing the energy consumption of ethernet with Adaptive Link Rate (ALR), *IEEE Trans. Computers* **57**(4), April 2008.

[17] IEEE P802.3az Task Force: Energy efficient ethernet, `http://ieee802.org/3/az/public/index.html`.

[18] Nedevschi, S., et al, Reducing network energy consumption via sleeping and rate adaptation, Proc. 5th USENIX Symposium on Networked Systems Design and Implementation, April 2008.

[19] Wang, Y., et al., Virtual routers on the move: Live router migration as a network-management primitive, ACM SIGCOMM 2008, August 2008.

[20] Namiki, S., et al., Dynamic optical path switching for ultra-low energy consumption and its enabling device technologies, SAINT 2008, July 2008.

[21] Baldi, M., et al., Time for a "greener" Internet, Greencomm'09, June 2009.

[22] Kurita, T., et al., A study for green network (2) – ECO switching, IEICE Technical Report, B-7-107, March 2009.

[23] Takagi, M., et al., 400 Gb/s hybrid optical switching demonstration combining multi-wavelength OPS and OCS with dynamic resource allocation, OFC/NFOEC 2009, March 2009.

[24] Wada, N., et al., Field demonstration of 1.28T bit/s/port, ultra-wide bandwidth colored optical packet switching with polarization independent high-speed switch and all-optical hierarchical label processing, ECOC 2007 PD3.1, 2007.

[25] Watabe, K. et al., Optical packet switching with contention resolution mechanism using PLZT switches, OFC/NFOEC 2008, OThA5, March 2008.

[26] Hidaka, M., et al., Applying low power consumption superconductive device technology to real-time waveform monitoring for photonic network, SAINT 2008, July 2008,

[27] Kameda, Y., et al., Design and demonstration of a 4×4 SFQ network switch prototype system and 10-Gbps bit-error-rate measurement, IEICE Trans Electron 2008 E91-C, 325-332, 2008.

[28] CREST, Ultra Low Power Consumption Information Technology, `http://www.ulp.jst.go.jp`.

[29] NTT, `http://www.ntt.co.jp/csr/2008report/ecology/activity01.html`.

[30] Ohara, S., et al., A prototype of a multi-core wireless sensor node for reducing power consumption, SAINT 2008, July 2008.

[31] Minami, M., et al., Solar biscuit: A battery-less wireless sensor network system for environmental monitoring applications. In *Proc. of the 2nd International Workshop on Networked Sensing Systems*, June 2005.

[32] Takiguchi, T., et al., A novel wireless wake-up mechanism for energy-efficient ubiquitous networks, GreenComm'09, June 2009.

[33] Zhou, X., et al., 32Tb/s (320×114Gb/s) PDMRZ8QAM Transmission over 580 km of SMF28 UltraLowLossFiber, OFC/NFOEC 2009, PDPB4, March 2009.

[34] Imaizumi, H., et al., Power saving mechanism based on simple moving average for 802.3ad link aggregation, GreenComm2, December 2009.

[35] Imaizumi, H., et al., Power saving technique based on simple moving average for multi-channel ethernet, FT3, OECC2009, July 2009.

4

Trend and Technology Development in Green Communications: KDDI's Technical Approach

Takeshi Mizuike and Keizo Sugiyama

Technology Development Center, KDDI R&D Laboratories, Inc., 3-10-10 Iidabashi, Chiyoda-ku, Tokyo 102-8460 Japan

Abstract

This paper provides an extensive survey on trend and promising technologies for power saving and ecology from the viewpoint of telecommunication operators. New technology development is presented with focus on the use of clean energy such as fuel cell and solar panel, improvement of power efficiency, and system monitoring and control for power saving. Activities of KDDI are then introduced for prototype development and field trials with regard to the use of fuel cell and solar cell for mobile handset, high efficiency power amplifier, power supply for cellular base station by solar cell and back-up battery and so forth.

Keywords: Clean energy, power saving, ecology, green ICT, telecommunication operator.

4.1 Introduction

Global warming is the most urgent and increasingly important issue, for which extensive efforts have been made in various industrial areas. In the field of Information and Communication Technology (ICT), a special attention has

R. Prasad et al. (Eds.), Towards Green ICT, 55–64.

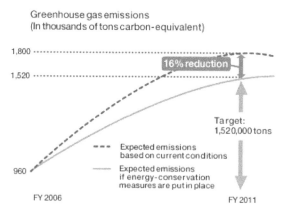

Figure 4.1 KDDI's targets for the reduction of greenhouse gas emissions (in thousands of tons carbon-equivalent).

been paid on power saving of ICT equipment as well as effective application of ICT technologies to reduce Greenhouse Gas (GHG) emissions.

It should be noted that telecom operators have been consuming a huge amount of electric power for 24 hour non-stop operation of communication equipment to maintain reliable telecommunication services. Recognizing a need for improvement, KDDI created the Environmental Charter in March 2003, which is a set of guidelines for environmental initiatives, and has announced an original action plan to prevent global warming [1]. The objective is to reduce estimated power consumption in 2011 by 16%. By reducing 520 million kWh, the total power consumption is estimated to be 2.74 billion kWh, which corresponds to 1.52 million tons of carbon equivalent as shown in Figure 4.1.

This chapter is intended to survey essential subjects for ecology and energy saving from the viewpoint of telecom operators and to introduce activities of KDDI on these technical subjects.

4.2 Essential Subjects in Ecology and Energy Saving for Telecom Operators

Telecommunication infrastructure has been highly developed to provide various services and contents to customers anytime, anywhere. As shown in Figure 4.2, advanced ICT technology can realize this seamlessly across fixed and mobile communication networks. In addition, boundary is disappearing

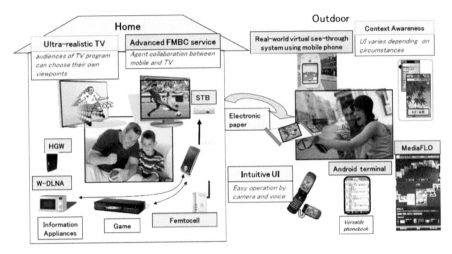

Figure 4.2 KDDI's FMBC service concept.

between telecommunication and broadcasting services such as one-segment terrestrial digital broadcasting to mobile handset and multi-channel broadcasting and Video on Demand (VoD) services through FTTH and CATV. Being the most essential element of such Fixed Mobile and Broadcasting Convergence (FMBC) services, cellular mobile networks have become indispensable life-line infrastructure. In Japan, more than 100 million handsets are in operation to cover 90% of households. The number of operational broadband access lines has exceeded 30 million, of which FTTH has about 50% of share [2].

Energy saving has become the most critical subject for telecom operators to maintain reliable telecommunication services in such highly advanced infrastructure. For instance, a huge number of base stations have been densely deployed to guarantee seamless cellular coverage. It has become necessary to lower the size and power consumption of base stations so as to reduce CAPEX and OPEX. Total power consumption is now no longer negligible for home gateway and set top box equipment that a large number of users install to enjoy broadband services. Electrical power requirement is soaring for operation of not only a large number of servers and network equipment but also air conditioning systems at data centers for so-called cloud computing and corporate network systems.

4.3 Overview of Promising Technologies

4.3.1 Clean Energy

Functions and applications of mobile handset have become so advanced that power requirements are always increasing. Flat-rate charging has encouraged users for longer use of handsets. Under limitations of small size and light weight, larger power supply has been required to guarantee longer battery life. Currently, Lithium-ion battery is most widely used. For example, a Lithium-ion battery of 2.16–3.6 Wh (0.6–1.0 Ah at 3.6 V) allows continuous operation of a 2-hour call. New applications such as mobile broadcasting reception, navigation and gaming will require further larger battery capacity.

A promising new technology for this usage is a compact fuel cell. Hydrogen type, however, is unsuitable for mobile handset applications due to its low energy density. Direct Methanol Fuel Cell (DMFC) is considered to be most promising. Energy density of DMFC including a fuel container is estimated to be about 2 Wh/g whereas that of Lithium-ion battery remains about 0.6 Wh/g.

Another attractive technology is solar cell as efficiency of energy conversion is rapidly improving. In application of solar cell to mobile base stations, it is necessary to overcome issues such as large area requirement, output power variation due to shadowing and rigid structure for stable operation under strong wind environment at roof top. Reduction of installation cost is a key factor for commercial use of this technology.

4.3.2 Power Saving for Base Station

Improvement of power efficiency (RF output power over DC input power) is a critical requirement of energy saving for power amplifiers. New technologies have become available for higher power efficiency. Doherty type is currently used most widely for high efficiency power amplifier with which Digital Pre-Distortion (DPD) is applied to compensate nonlinear distortion. Development is underway for new technologies such as switching amplifier of E and F classes and envelope tracking which controls DC power supply for the power amplifier. Figure 4.3 shows technology trends of high efficiency power amplifier [3]. Further improvement of 10% is expected by these new technologies.

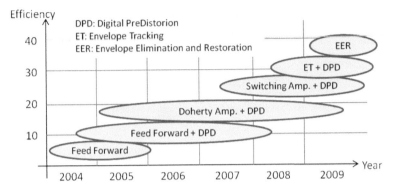

Figure 4.3 Technology trends of high efficiency power amplifier.

4.3.3 System Monitoring and Control

Power supply of cellular base stations consists of a DC power converter for commercially supplied AC power from an electric power company and a back-up battery. It is requested by law to prepare battery power supply for back-up of power failure of at least three hours. It should be noted that electric power demand is heavier during day time, when power supply is more dependent on thermal power generation which tends to emit more GHG. It is then effective to control power supply in such a way as to charge up back-up battery during night time and to use the back-up battery during day time in combination with commercial power supply. Use of commercial power supply during night time is advantageous due to lower electric power price as well as less GHG emission.

Power saving by monitoring and control is effective in various situations. At data centers, it is possible by virtualization technologies to change the number of operational servers flexibly in proportion to processing load. For home appliances, stand-by power consumption may be reduced by monitoring operational status of equipment to shutdown unused appliances.

4.4 KDDI'S Activities for Green ICT

Increasing power consumption of ICT has become a big concern as the use of telecommunication services continues to grow. KDDI has been actively involved in development and installation of equipment with lower power consumption, efficient system operation for power saving and use of clean

(a) Fuel cell (attachment type)   (b) Fuel cell (built-in type)   (c) Solar cell

Figure 4.4 Fuel cell and solar cell for mobile handset.

power supply. This section introduces example cases of development and trial that KDDI recently conducted.

4.4.1 Use of Fuel Cell and Solar Cell for Mobile Handset

KDDI first developed a prototype fuel cell as an external attachment of a mobile handset as illustrated in Figure 4.4a. This external type of fuel cell is capable of 1 W power supply. KDDI has recently succeeded in development of a built-in type fuel cell for a mobile handset as shown in Figure 4.4b. In the case of the built-in type, the output power of the fuel cell is about 300 mW. It is possible to refill the fuel from a cartridge containing methanol.

Another example is a mobile handset which is equipped with a solar cell panel on the external surface of the handset body. Figure 4.4c shows a photo of a commercially available product. Ten-minute charging by solar power generation allows a cell phone operation of a one-minute call or a two-hour call waiting.

4.4.2 High Efficiency Power Amplifier

Gallium Nitride (GaN) devices are very promising for excellent performance at high frequency bands. Compared with conventional devices such as Silicon and Gallium Arsenide, GaN device allows high voltage and high frequency operation in so-called High Electron Mobility Transistor (HEMT) structure. GaN devices are currently about three times as expensive as conventional products. As the market size grows, commercial use of GaN devices is expected to start in very near future when the cost is successfully reduced. To take maximum advantage of GaN device, design of high efficiency power ampli-

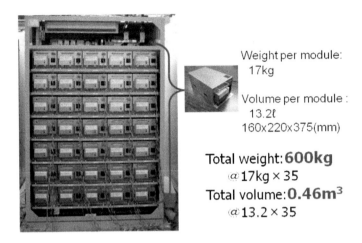

Weight per module:
17kg

Volume per module :
13.2ℓ
160x220x375(mm)

Total weight: **600kg**
ⓐ17kg × 35
Total volume: **0.46m³**
ⓐ13.2 × 35

Figure 4.5 1750 Ah Li-ion battery system.

fier was optimized. Improvement was also attempted for DPD technology to compensate non-linear distortion arising from the power amplifier.

A prototype power amplifier was developed for WiMAX applications to achieve about 30% of power efficiency, which is twice as high as conventional products for 25 W output power and operational frequency band at 2.5 GHz. Excellent performance of this new technology was verified for commercial applications [4].

4.4.3 Power Supply for Cellular Base Station by Solar Cell and Back-up Battery

Base stations of cellular network are equipped with back-up power supply to guarantee high operational availability. So far, lead battery was commonly used as back-up power supply. Recently, the Lithium-ion battery is drawing attention due to its excellent performance, such as compact equipment size, light weight, less environmental pollution and excellent electrical perform- ance of charging and discharging. KDDI has developed a prototype back-up power supply system using a Lithium-ion battery as shown in Figure 4.5. This prototype Lithium-ion battery system which is capable of providing 1750 Ah at DC 27 V, is currently under field testing at an operational base station.

KDDI is currently conducting another field trial for combined operation of solar cell power generation and back-up power supply as shown in Fig- ure 4.6. During night time, the back-up battery system is fully charged by

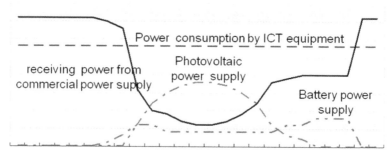

Figure 4.6 Combined operation of solar cell power generation and back-up power supply.

commercial power supplied by an electric power company. During day time, the solar cell is used to the maximum extent possible in combination with supplement power supply from the back-up battery so that the use of commercial power supplied by the electric power company may be minimized.

It should here be noted that price of the commercial power during night time is generally reduced. In addition, electric power during night time is generated by power plant with less GHGs such as nuclear power plant and hydropower plant. It is expected that the combined operation of solar cell and back-up battery contributes to not only cost efficient operation but also reduction of GHGs.

4.4.4 Monitoring and Control of Home Appliances by Cellular Handset

In 2003, KDDI developed a prototype home network that consisted of a set of wireless nodes called ubiquitous node [5]. Each ubiquitous node provides interface function for control of home appliances and environment sensing. One of the ubiquitous nodes functions as a gateway for connection to Internet. As shown in Figure 4.7, this network can collect sensing information such as room temperature, humidity, brightness and human motion to inform a mobile handset user via Internet and cellular network.

In this network, it is also possible for a mobile handset to control home appliances through cellular network and Internet. The users can easily utilize functionality for home security and remote control of home appliances. This prototype network has automatic configuration capability to set up the gateway node and a set of ubiquitous nodes for ubiquitous wireless home network. Effective use of this home network will contribute to power saving of home appliances.

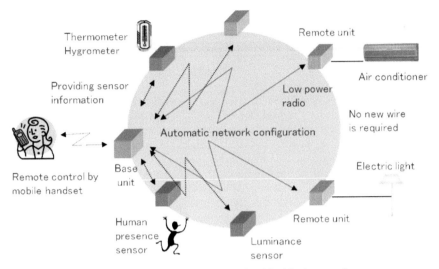

Figure 4.7 Prototype home network with ubiquitous node.

4.4.5 Recycling of Mobile Handset

The mobile handset market is so active that new handset models are released very frequently. Every time a new model is released, a number of subscribers replace their old handsets by brand new products. Recycling of used mobile handsets has become a serious problem to avoid wasting precious materials such as rare metals in the used mobile handsets. During the process of recycling used mobile handsets, manual disassembly was introduced to collect precious materials as much as possible. As a result of this manual effort, the reuse percentage of precious materials is getting closer to 100%. Recycling has also been started at mobile handset retail stores to collect used user's manual documents.

4.5 Conclusions

In this chapter, various activities for energy saving of ICT equipment were introduced with focus on example cases in KDDI. In the future, equalization of power load by ICT will become effective by real time monitoring of power consumption and visualizing energy flow. For this purpose, it is also effective to make use of power supply by distributed power generation using natural energy and electric power storage. Application of ICT is expected to contribute to change of life style as well as power saving.

References

[1] KDDI's CSR, Environmental Conservation, `http://www.kddi.com/english/corporate/kddi/csr/environment/environment/index.html`.

[2] Hishinuma, H., ICT policy in Japan – Broadband and mobile. Paper presented at High Level Seminar on ICT, April 2009.

[3] Kanbe, A. et al., New architecture of envelope tracking power amplifier for base station. Paper presented at the 21st Workshop on Circuits and Systems, Karuizawa, April 2008.

[4] KDDI News Release, KDDI and Fujitsu develop practical-use high-efficiency amplifier for mobile WiMAX, `http://www.kddi.com/english/corporate/news_release/2007/0302/index.html`.

[5] Yoshihara, K., Motegi, S., and Horiuchi, H., Design and implementation of "kubit" for sensing and control ubiquitous applications. In *Proceedings of 3rd IEEE Conference on Pervasive Computing and Communications Workshops (PerCom 2005 Workshops)*, pp. 189–193, March 2005.

5

Role of ICT in Positively Impacting the Environment

Kishore Ramareddy[1], Parag Pruthi[1] and Ramjee Prasad[2]

[1]*NIKSUN Inc., Princeton, NJ 08540, USA*
[2]*Department of Electronic Systems, Aalborg University, 9220 Aalborg, Denmark*

Abstract

Modern societies dependence on energy is increasing at a rapid rate. Several studies have shown that the annual global demand for power consumption is increasing at the rate of 16–20%, which equates for a doubling of the demand every 4–5 years [1, 2].

This growing demand for energy comes at a cost to our environment due to the increased green house gas emissions generated from the various forms of energy presently being utilized. Not only is there an impact on the global climate, but we are also beginning to see its impact on various life forms. Several studies [3] have revealed that, with the increase in the temperature due global warming, certain species have started to breed and migrate earlier than expected. Other studies have found that the species have shifted closer to poles or moved to higher altitudes indicating that plants and animals are occupying areas that were previously considered too cold for survival.

However, not everyone is convinced that the increased carbon dioxide emissions have much impact on our environment. The observations of global climate change are being attributed to natural drifts in climate and it is suggested that deviations in temperature are to be expected.

Regardless of ones belief in global climate change or not, continuing to consume energy resources as we have been doing for the last century is not sustainable. Without regard to the impacts on the environment, we need to

R. Prasad et al. (Eds.), Towards Green ICT, 65–74.

develop a culture of economical and sustainable energy. In this greener future, ICT can play an important role. We will show that ICT can have a dramatic and positive impact on the environment.

Keywords: Climate change, green house gasses, green communications, sustainable energy, ICT and the environment.

5.1 Introduction

According to the American Energy Information (EIA) and to the International Energy Agency (IEA), the worldwide energy consumption will on average continue to increase by 2% per year. This leads to a doubling of the energy consumption every 35 years. EIA estimated that in the year 2007, 86.4% of the energy was generated using the fossil fuels. Clearly such consumption cannot be sustained, as the availability of fossil fuels is limited. In addition to the potentially unsustainable rate of increase, the byproducts of green house gases are projected to have a significant impact on our environment [16].

It is estimated that ICT contribution is around 2% of the world's green house emissions [4]. Even though the overall contribution of the ICT appears to be low, it is growing at an accelerated rate. Looking more specifically at this industry, studies have shown that the demand for energy in data centers is exceeding at an exponential rate. During the early part of this century, a rate of energy consumption have doubled in only five years and is expected to double again 2011 [5]. Worldwide in 2000, the data centers made up about 14.1 million installed servers, whereas by 2005 the number of servers in data centers climbed to about 27.3 million. Data center energy consumption is not only a result of the servers, but up to half of the energy consumed may be from computer operations, the other half being attributed to cooling systems or uninterrupted power supplies [17]. In fact the focus on energy consumed by data centers has become so high, that it is one of the top concerns for large technology companies such as Google, Vodafone, Microsoft, etc. In fact energy consumption is the single biggest challenge at Google.

In addition to data centers, general consumption of electricity by things just idling and not producing anything useful, such as power adapters plugged into the wall but not charging, idling computers, unused network disk drives, etc., is adding significantly to wasted energy.

As another data point in evaluating the efficiency of ICT systems, the energy usage of cell sites in the wireless industry can illustrate the need for higher efficiencies. A traditional cellular site requires about 1500 Watts of power whereas the transmitted power is roughly 60 Watts [6]. Therefore, there is a lot of ancillary power wastage that can be targeted to make a cellular site more efficient. Obviously the wireless devices themselves have been and are constantly being optimized for longer operation and standby times.

Given that the ICT is itself not a major contributor to green house gasses, how can the ICT help in positively impacting the environment? First, the ICT industry seems to be aware of the growing challenges and various initiatives are underway. We shall discuss some of those in this chapter. Secondly, the biggest contribution that ICT can have is in making other industries more efficient. Developing sensors and technology that can adapt the energy consumption of household devices and environment, office and industrial work environments, factories, power generation systems, transportation systems, etc., can significantly impact the environment in a positive way.

5.2 Motivation

Whilst there is significant awareness of the impacts of global warming, there are still many who hold the belief that the data presented by the environmentalists is a phase and is part of the earth's natural cycle [7]. They contend that the earth is a self-regulating system and that it will naturally correct any irregularities that may crop up from time to time.

Although several researchers have developed sophisticated models for earth's climate, one cannot be sure whether the observations of current climatic changes are subject to self-regulation during a period of time that does not have impact on the various forms of life on earth. In fact since we do not have a robust and tested model for the earth's climate, one does not know whether the system is currently in a locally stable point in a chaotic system. If the climate is indeed described by a chaotic system, which generally is the rule for complex natural systems, then it could be likely that a small change in the environmental conditions could trigger very large changes in the environment. For a simple example of how this can happen, one needs to only study the very simple logistic map, which is frequently used to describe population dynamics in a world with finite resources (or food). The simple equation of the logistic map is $p_{n+1} = r \cdot p_n(1 - p_n)$ where p_n is the population at time n and r is the rate of population growth. Figure 5.1 shows iterates of the logistic map for various values of r.

As can be seen, for values of $r < 3$, the population converges to a stable point, but at $r = 3$ a bifurcation occurs, i.e. p could be at one of two values. As we further increase r, we see that the value of p could oscillate between

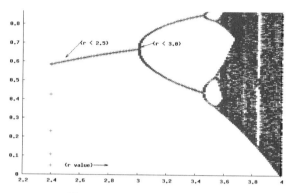

Figure 5.1 Bifurcation diagram.

many different values. Indeed at some of the "islands" we can see that *p* could remain locally stable for a while, but that it could suddenly changes its value to another value far from the previous value. Such systems are more common in nature than one can imagine. It is possible that the earth's climate is governed by a similar system. If that is the case, then a slight change in the environmental conditions could trigger the earth's climate to suddenly change to a significantly different locally stable region. Obviously this could have a dramatic impact on all life forms.

The supporters of the green movement are not only concerned about local changes in climate but are also pointing to the potential long term harmful effects that could arise. Then there are those who seem to be skeptical, as local observations in climate change sometimes seem to suggest that the climate is not getting warmer. Figure 5.2 shows the average global warming over the last century [8]. It generally shows a warming trend when looking at the five year moving average.

Typically a chaotic system can exhibit similar behavior. As shown by the logistic map, local oscillations may not display much change over a short period of time (years to centuries in the global context) but the system could be slowly moving into a region of instability. Once it reaches a critical threshold, the system could dramatically shift into a different locally stable region and in that region the system could have devastating effect on life forms.

Therefore, given that there is a limited supply of fossil fuels and our understanding of global climate and its impact on life forms is improving, it only makes sense to use alternate sources or renewal energy and to develop the science of efficient power use and consumption.

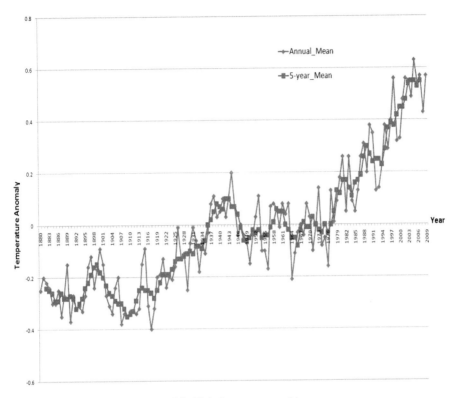

Figure 5.2 Global temperatures history.

5.3 ICT's Role

5.3.1 Reducing Power Consumption of Electronic Devices

In general, the ICT can set targets for reducing power consumption of various electronic devices.

As an example, the energy cost estimate of PC is roughly $55–70 per year as a result of it simply idling. It is estimated that there are roughly 100 million computers [9] running windows XP that results in about $5 to $7 billion spent on energy on an annual basis for idling computers. This is equivalent to the CO_2 emissions for one year produced by every household and industry for a country the size of Ireland! Significant advances can be made in this area as shown by Sony in its PS3 Slim's power consumption over the original PS3; while idling the original PS3 consumes about 1.22 Watts of energy whereas the PS3 slim uses only 30% of that or about 0.36 Watts of energy [10]. And in general, the PS3 slim generally uses half the power of the original PS3! This goes to show that by using modern power saving techniques and electronics, power consumption could dramatically be reduced directly by the actions of the ICT community.

Methodologies to reduce power consumption include use of low power electronics such as OLED displays, power adapting sub-components, etc. In the communications industry a variety of techniques are being developed such as:

- Schedule-Based Power over Ethernet (PoE).
- Auto turn off of devices not being used (e.g. Ethernet ports, WLAN devices, etc.).
- Inactivity timers to power down hard drives and other mechanical devices when not in use.
- Self regulating fans that can lower their speeds to lower temperature scenarios.
- Automatic power adjustment based on cable length and/or communicating clients on wireless zones.

In a standard workday scenario where computers are used for 10 hours a day and powered off 14 hours a day, and connected to the switch using 20 meter cables, D-Link estimates that its products can save up to 66% power used for each system by using some of these green technologies [17].

5.3.2 Reducing Power Consumption in Data Centers

As we have stated earlier, data centers power consumption is growing exponentially. Not only do the network and computing devices in the data center need to be more power efficient but also the power and cooling systems need to be more efficient. Up to half the power consumed in a data center can be from sources other than the network and servers. A survey sponsored by 1E and the Alliance to Save Energy found that one in six servers – about 4.7 million worldwide – are doing nothing useful, costing businesses as much as $25 billion a year [11]. Therefore, virtualization, load balancing, and more efficient use of resources can significantly improve utilization of such servers. Note that there may be peak demands, however if the peak demands follow a pattern, such as follow the sun, then an appropriate connectivity and access via a global information grid can help companies utilize such resources in more efficient ways.

Generally, the following methods can be utilized to improve data center efficiency:

- Appropriate use of cooling such as hot isle, cold isle techniques can substantially reduce power consumption necessary to cool a data center. Cooling efficiency is increased by approximately 20% [12].
- Efficient utilization of servers by using techniques such as virtualization and making them available for use as needed by a global network.
- Use of DC instead of AC to increase efficiency by 20% [13].
- Utilizing technologies that can run "hotter" so that the need for cooling can be reduced.
- Using direct heat exchange between data centers and water bodies or other communities that need the heat such as swimming pools, community centers, etc., in cold climate areas. Recently IBM and GIB announced a data center at a former military bunker in Uitikon, Switzerland, where the heat generated from this data center will be used to heat a public swimming pool in the town [14].
- Developing novel ideas such as floating data centers on the ocean as per a patent application by Google [15].

5.3.3 Proper Use of Packaging

An indirect contribution to our environment is the improper use of packaging materials. Often packaging materials are overlooked and create not only waste material that is hard to recycle, but may itself be a consumer of expensive energy to produce. As such, the ICT industry can adopt using more efficient methods of packaging and disposal of those materials.

Some of the efficient packaging methods include:

- Shrinking the packaging size, there by reduce the amount of cushioning materials.
- Reduce package weight by using air-filled cushion technology.
- Use renewable, non-toxic and 100% recycled materials (e.g. reprocessed paper, vegetable based inks) for packaging.
- Shrink-wrapping the box, so that there is no need of plastic bags for individual accessories.

It is estimated that Dell will save $8 million or more by using these methods while eliminating some 20 million pounds of packaging materials over the next four years [18].

5.3.4 Developing Technologies That Can Minimize Green House Gas Emissions

The greatest contribution ICT can make is to develop techniques, especially sensing techniques that can regulate energy use to minimize green house gas emissions.

Developing sensing and adapting technologies for residential and commercial environments can have significant impact on the overall reduction of green house gas emissions. Typically, houses and commercial buildings do not adapt very well to time of day changes in weather or to inclement weather conditions. Self-regulating buildings that are "aware" of the changes in weather ahead can better adjust the climate control systems in a more efficient way. Methods to transfer solar energy from the sun facing side of the building to the other sides of the house can also reduce the energy requirements.

There are many such developments that ICT can make which will reduce the overall green house emissions. Some of them are:

- Climate control systems that adapt to the weather forecast.
- Human sensing technologies that adapt the climate, lighting, refrigeration, and other devices (e.g. turn off the microwave, coffee pot, etc., if no one is around the kitchen).
- Thermal regulators that move heat from one area to the other. Example, if one down the temperature on an upper level and no one is in the basement which is generally cooler, then transfer heat from the upper floors to the basement.
- Power sockets that adapt to energy consumption of devices. Example, if one has leak current being drawn by chargers that are not connected to a device, then power off that electrical socket all together.
- Self-adjusting blinds and windows that can adapt to the current environment to provide the most efficient movement of energy, etc.

ICT can also build technologies that can modify user behavior towards a more efficient energy system. By devising technologies that encourage less transportation of materials and more efficient use of shared resources, ICT can have a great impact on the use of energy in transportation. For example, by developing ways in which electronic payments can be made one can not only reduce paper bills, tickets, etc. but can also encourage use of transports and facilities that provide such facilities. Use of better virtual reality systems that can bring communities of users together can have a greater impact on reducing the overall need for transportation. With such systems, the need to visit physical stores to "look" at goods can be greatly reduced.

5.4 Conclusions

The ICT can contribute in many ways to reduce green house gas emissions and positively impact the environment. First and foremost, the ICT can develop technologies that are efficient, require less power, are friendlier to produce and dispose off and are more efficient to package and transport. In particular, the ICT can have a positive impact on energy use in data centers and personal computing devices.

However, given the relatively small percentage of the ICT's contribution to green house gas emissions, the ICT can develop technologies that can help reduce green house gas emissions in other sectors such as industrial processes, power stations, residential and commercial, transportation, etc. By developing technologies such as efficient power consumption, efficient data centers, green packaging, adaptive climate control mechanisms, etc., the ICT can help reduce dramatically the green house gasses emitted by these other sectors.

In the end, the debate whether climate change is real or not is irrelevant. Our dependence on non-renewal sources of energy must be reduced and even when we have developed alternate, locally produced and renewable energies, the waste of these energies will have an impact somewhere down the road. Developing efficient and bio friendly techniques can only have a positive effect on generations to come.

References

[1] Rate of increase in power consumption, references 1–3, e.g. TU Dresden, IPCC.
[2] Intergovernmental Panel on Climate Change. Climate Change and Water, Effects of global warming already being felt on plants and animals worldwide, *Science Daily*, 2003, http://www.sciencedaily.com/releases/2003/01/030101222546.htm.
[3] Gordon, J., WEF (Switzerland), World Economic Forum work on ICTs and climate change, 2006, http://www.itu.int/dms_pub/itu-t/oth/06/0F/T060F0060080013PDFE.pdf.
[4] Report to Congress on Server and Data Center Energy Efficiency Public Law 109-431, U.S. Environmental Protection Agency, August 2007.
[5] Key best practices, Increase your data center energy efficiency, U.S. Environmental Protection Agency, http://www1.eere.energy.gov/femp/pdfs/data_center_qsguide.pdf.

[6] Hérault, L., Green wireless communications, `http://leti.congres-scientifique.com/annualreview2009/AnnualReview/AR09_session2_Herault.pdf?PHPSESSID=067f032961804aa8077c2fefa71762f8`.

[7] Essenhigh, R.H., Does CO_2 really drive global warming, *Chemical Innovation*, Vol. 31, No. 5, pp. 44–46, May 2001.

[8] NASA GISS Surface Temperature Analysis, `http://data.giss.nasa.gov/gistemp/graphs/`.

[9] Daub, T., Microsoft could save 45 million tons of CO_2 emissions with a few lines of computer code, `http://blog.foreignpolicy.com/posts/2006/11/15/microsoft_could_save_45_million_tons_of_co2_emissions_with_a_few_lines_of_computer_`.

[10] PS3 Slim uses half the power of PS3, CNET, `http://news.cnet.com/8301-17938-105-10318727-1.html`.

[11] Burt, J., Unused servers cost business $25B annual: Study, `http://www.eweek.com/c/a/Green-IT/Unused-Servers-Cost-Businesses-25B-Annually-Study-582507/`.

[12] Using curtains to manage data center airflow, `http://www.thehotaisle.com/2008/04/18/being-smart/`.

[13] Tschudi, W., DC power for improved data center efficiency, Lawrence Berkley Laboratory, March 2008.

[14] IBM builds green data center for GIB-services; Innovative technology to heat local public swimming pool, `http://www-03.ibm.com/press/us/en/pressrelease/23797.wss`.

[15] Water based data center, `http://appft1.uspto.gov/netacgi/nph-Parser?Sect1=PTO1&Sect2=HITOFF&d=PG01&p=1&u=%2Fnetahtml%2FPTO%2Fsrchnum.html&r=1&f=G&l=50&s1=%2220080209234%22.PGNR.&OS=DN/20080209234&RS=DN/20080209234`.

[16] The expected global impact consequences of greenhouse gases and climate change, `http://www.azocleantech.com/details.asp?ArticleID=102`.

[17] D-Link Green Technology, `http://www.dabs.com/learn-more/networking-and-communications/d-link-green-technology-8267.html`.

[18] Dell looks to green packaging to save, `http://www.pcworld.com/businesscenter/article/156169/dell_looks_to_green_packaging_to_save.html`.

6

Green Mobile

Knud Erik Skouby and Iwona Windekilde

*Center for Communication, Media and Information Technologies (CMI),
Copenhagen Institute of Technology/Aalborg University, 2750 Ballerup, Denmark*

Abstract

The concept of 'Green Mobile' is getting increasing attention as part of the concern regarding climate change and environmental sustainability issues. The wireless/ mobile sector was, however, a late comer in the debate. The discussion started with an ITU document in 1994 and with a unilateral focus on the positive role of introducing ICT in other sectors, using

- the latest telecommunication and information technologies to monitor pollution, wildlife studies, forestry development, and others;
- telecommunication technology to reduce paperwork, which ultimately saves forests.

Only from 2007/2008 a more balanced and advanced view has found its way into the debate on ICT and sustainable development. The impact of ICT is now seen as three-fold:

- *Directly*, by reducing the ICT sector's own energy requirements and 'pollution effects'.
- *Indirectly*, through using ICTs for 'virtualization' (video-conferences, online activities, etc.).
- *'Systemic'*, by providing the technology to implement and monitor resource reductions in other sectors of the economy.

From now on the interest is not only on the positive effect of ICT on other sectors, but also on the activities within the sector itself thereby demanding actions within the sector.

R. Prasad et al. (Eds.), Towards Green ICT, 75–86.

The chapter discusses this development, the driving forces and, e.g., the resulting new business models developed under increasing pressure from customers, shareholders and proposed legislative changes to improve their environmental credentials.

Keywords: Green mobile, ICT, green business model, energy efficiency, green initiatives.

6.1 Introduction

The ICT industry has a very significant role to play in reducing greenhouse gas emissions. Many sources has assessed that ICT is responsible for rather modest 2% of the world's total emission of CO_2, but contains solutions to reduce the remaining 98% of CO_2 emissions significantly [1].

The increasing spread of ICT has of course negative environmental consequences, but is also part of the solution to better living and a cleaner environment. ICT has the potential to make direct and indirect contributions to reduce the environmental impacts on other areas of the economy, for instance through the reduction of transport needs, reduction of paper use, and the intelligent management of energy consumption. It has been estimated that through enabling other sectors to reduce their emissions, the information and communications technology industry could reduce global emissions by as much as 15% by 2020 – a volume of CO_2 more than five times its own footprint [2].

Besides the energy consumption aspect, ICT equipment often contains chemicals harmful to health and to the environment. According to the International Telecommunication Union about 60% of the world's population now have a mobile phone, with approximately 4.6 billion mobile-phone subscriptions by the end of 2009 [3]. Moreover, mobile terminals and equipment is regularly replaced by users due to new functionality or fashion trends and thus large amounts of devices need to be handled. Therefore it is important to minimalize environmental impact throughout the whole life cycle of mobile ICT equipment: from development, over production and usage to disposal.

In this paper the direct, indirect and systemic impact of ICT on environment is discussed. First an overview of Green IT concept is given, then important issues related to regulation are presented, finally, the environmental impact of mobile market is analysed and the specific cases of green mobile solutions implementation are described. The paper also outlines green business models as well as awareness and demand for green mobile solutions.

The paper ends with a conclusion including a suggestion for a framework promoting 'Green Mobile' solutions.

6.2 Green Mobile Concept

Overall Green IT is research in and use of ICT in an efficient and environmentally friendly manner. The realization of the Green Concept is targeting to reduce costs, increase productivity, and improve system performance, while minimizing the negative environmental effects of ICT.

The interest is not only on the positive effect of ICT on other sectors, but also on the activities within the sector itself including the environmental impacts of mobile communications devices in terms of the resources and energy they use and also on more environmentally friendly utilization of ICT.

Within ICT, 'Green Mobile' is getting increasing attention as 'Green' is more specifically defined as:

1. meeting the needs of present generations without compromising the ability of future generations to meet their needs,
2. pollution prevention at the end of a product's use,
3. product stewardship to minimize the environmental footprint during use, and
4. use of technologies to reduce the use of polluting materials.

The need for Green Mobile seems evident and new initiatives are springing up constantly, but the impact so far has been tiny. Obviously, the initiatives are very recent and results may show, but maybe the efforts do not match the needs and potentials. This is discussed below.

6.3 Regulatory Drivers for Green Mobile Solutions

In the context of the Kyoto Protocol and the EC's commitment to cut CO_2 emissions, the improvement of energy efficiency, rational use of energy, and the promotion of new and renewable energy sources are considered as cornerstones of the EU energy policy objectives. EU support is emphasized in the Commission Green Paper towards a European Strategy for Energy Supply Security which highlights the central role of energy efficiency for increasing the security of supply and reducing greenhouse gasses emissions [4].

This is further developed in the European Climate Change Programme (ECCP) [5] which highlights the large cost-effective potential for improving energy efficiency of end-use equipment.

Examples of regulation relevant for mobile phones are: the EC Directive on Waste Electrical and Electronic Equipment (WEEE) [6], the Reduction of Hazardous Substances (RoHS) [7] directive; and the Energy Using Products (EUP) directive [8]. These directives require companies to generally implement policies relating to the purchase, use and disposal of ICT equipment, and also provide solutions to consumers for disposing their electrical and electronic waste without any additional costs.

Considering the complex nature of mobile phones, a variety of products available, the rapid technology development, and the different stakeholders involved in it could be very valuable to use not only regulation but also market based instruments such as voluntary agreements amongst companies.

Recently the EU has taken a significant step towards convincing mobile phone producers to incorporate voluntary agreements towards harmonization of chargers. The move announced in June 2009 followed a request from the Commission for a voluntary commitment, negating the need for legislation. Consumers will benefit from lower prices, new phones will not have to come with new chargers thanks to this reusability, preventing old chargers becoming electronic waste [9].

The EU planes to use ICTs to monitor, optimize, and reduce the energy consumed in various sectors, e.g. construction, heating, cooling and lighting of buildings. Therefore it is expected that by 2012 companies will be forced to take a first step towards disclosing the extent of their CO_2 emissions.

To further stimulate reduction or elimination of materials of concern, several policy options could be implemented like fiscal tools and mandatory warnings on products that contain such materials.

In order to create consumer demand for products free of these materials more sophisticated information and policy tools need to be used to stimulate the environmental improvements in mobile phone products.

6.4 The Growth of the Mobile Market and Its Environmental Impact

6.4.1 Direct Impact

6.4.1.1 Terminals

The most important life-cycle environmental issues for mobile phones are:

- carbon emission associated with handset production process,
- energy consumption during the usage,

- presence of some materials of concern in phones,
- collection of unwanted phones and their recycling.

Issues related to size and weight of mobile phones are already optimized as they are driven by business/customer requirements. Issues regarding energy consumption, more environmentally friendly production, presence of harmful materials, and recycling of mobile phones are, however, still open in relation to the environmental improvements.

A major contribution of mobile (and ICTs) to climate change comes from the growing number of user terminals/ devices. Mobile phone use has grown rapidly with 1.22 billion handset sold in 2008 [10]. Power consumption has grown not only with the number of terminals, but also as the terminals become more advanced; 3G phones, e.g., need more power than 2G ones (Internet access, digital signal processing, polyphonic ringtones etc.).

In order to save energy, device manufacturers are working on increasing energy efficiency in the handset, increasing energy efficiency of the charger, using solar-power for handsets or for charging, etc.

According to the GSMA Association, the main impact associated with day to day mobile phone use is the power used during the charging process [11].

Recently more and more companies have decided to ensure more environmentally friendly production and to implement recycling programs in their green initiatives, e.g.:

- Motorola Renew is a basic terminal made out of recycled water bottles – no bluetooth, no GPS, no video player, no camera, no Wi-Fi.
- Nokia's 'Remade' – concept phone is built almost entirely out of re-cycled materials, including aluminum cans, plastics from drink bottles, and old car tires.
- Nokia 3110 Evolve is an eco-friendly and energy-efficient mobile phone made from over 50% renewable material. It is also packaged in 60% recycled content and comes with an efficient charger that uses 94% less energy than the old ones used.
- Samsung Blue Earth (launched in 2009) – is a touch screen solar-cell mobile phone made from recycled plastic culled from plastic water bottles. Both handset and charger are free of toxic materials. A full solar charge – 10 to 14 hours – provides power for four hours of talk time.
- Digicel, phone operator which operate in developing countries across the Caribbean, Central America and the South Pacific – launched the Coral-200-Solar, claimed to be the world's first ultra-low-cost solar-powered

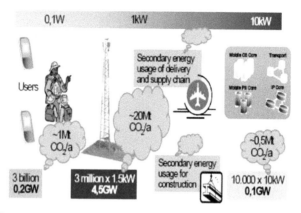

Biggest part of energy consumption in mobile base stations

Figure 6.1 Energy consumption in mobile [12]. Source: Nokia Siemens Networks.

mobile phone. Coral-200-Solar has an integrated solar charger built into the phone,

- T-Mobile USA has announced the introduction of Green Perks, a new application promising exclusive discounts on environmentally conscious products and services.
- Safaricom, Kenyan mobile network operator, launches Solar-Powered Mobile Phone (2009). Simuya Solar has been manufactured under a partnership with ZTE, the handset has been made by them from recycled materials and it possesses an in-built solar panel. Safaricom has got more than 60 Base Transmission Stations (BTSs) that are being operated on renewable energy sources wind and solar-driven turbines in various parts of the country.
- Sony Ericsson plans to make all phones green.

6.4.1.2 The Networks

From the perspective of a mobile system, the biggest part of energy consumption is in mobile base stations (Figure 6.1), amounting to as much as 80%.

In the future, mobile towers could be self sufficient if wind turbines and solar system are used to power them. Currently mobile towers are powered by diesel generators, which are not only harmful for the environment but also very expensive to operate. In Africa over 30 million litres of diesel per annum is consumed powering base stations (an average of 18,000 litres per base station per year). In this context wind and solar power is a feasible and cost effective alternative to using fuel generators at places where the main grid connection is not available.

ABI Research is forecasting that in 2009 more than 800,000 base stations will utilize wind or solar energy and there is a potential for a 30% reduction of carbon emission [13].

Greater emphasis on energy savings in the whole economy, together with new legislation mandating the use of renewable energy, will drive the adoption of Green IT solutions in the network infrastructure. As an example:

- Telenor (Pakistan) has announced a new energy efficient GSM network that will reduce its energy consumption by 50%.
- >100,000 Huawei green base stations have been deployed reducing CO_2 exhaust by 170,000 tons – equal to the CO_2 exhaust volume of 70,000 Chinese in one year.
- Global auditors, KPMG is replacing a fixed LAN with a Wi-Fi solution at its new Global Headquarter in Amsterdam saving US$2 million on its network building and reducing its annual OPEX by an estimated US$760,000.

6.4.2 Indirect Impact

The green role of ICT not only includes emission reduction and energy savings within ICT sector, but also very broadly encompasses the adoption of ICT technologies to influence and transform the way the society works and the way people behave. Illustrative examples include:

- Green IT services in the Asia-Pacific region (even excluding Japan) will grow to a US$2 billion opportunity by 2011 (Springboard Research).
- M-commerce: it is expected that by the end of 2009 74,4 million people will be using m-commerce worldwide; this number will double by the end of 2012. This will reduce person-transport and paper work.
- E-working: close to 100 million workers are expected to e-work either full-time or part-time by 2010. By implementing e-work companies will be able save money, enhance distributed work through e-collaboration, and manage the virtual workforce through effective communication
- Videoconferencing: There is an increased awareness and emphasis on reducing carbon footprint by cutting back on travels. This is one of the factors positively impacting the demand for videoconferencing.

- Globalization and an increasing number of remote workers is another factor behind the demand for new ICT solutions. Remote workers can communicate and collaborate regardless of localization, time, network or devices, also reducing travels and transportation.

6.4.3 Systemic Impact

Travel cost and time savings mentioned above have been key drivers for new ICT solution, but they are not the only factors influencing fast ICT adoptions in other sectors of the economy.

Comprehensive Green ICT solutions offer numerous benefits for economy as a whole. It increases employee satisfaction, comfort, enable new knowledge, save resources, enhance safety and security, protect health, improve food and water quality, improve productivity.

ICT applications and wireless sensor networks provide not only new ways to communicate and transfer information, but also have a significant impact on the environment, e.g. smart buildings, logistics and transportation, environmental monitoring, security and surveillance, health care, animal tracking and precision agriculture, and smart grids & energy control systems).

More specifically, there are spill over effects on the rest of the economy as ICT diffusion leads to innovation and efficiency gains in other sectors.

6.5 'Going Green' Business Model

Implementing green business model means for companies being efficient and not wasting money, resources and time, particularly as wholesale energy prices increased considerably during the last few years. In the long term, energy prices are expected to rise, both as a consequence of the increased global demand and of political requirements for reducing CO_2 emissions. As energy supply is becoming an increasingly important cost factor for companies, it is in this area in particular that green ICT offers significant savings potential, through more efficient hardware and the intelligent use of energy.

Green ICT has, however, a positive impact not only on costs, but also on the company's revenue. According to a recent study by the market research company McKinsey, 21% of end customers already deliberately choose products from companies acting in a sustainable and environmentally-aware manner and accept a higher price for this. A further 13% are also prepared to "pay" for this commitment to the environment, but have not yet put this willingness into action. The noticeable environmentally-aware behaviour of the company thus represents an additional – and perhaps decisive – selling point towards end customers and thus also opens up new customer groups [14].

Moreover there are good financial results to be obtained from Green ICT initiatives. Renewable energy solutions can in certain geographical areas simultaneously expand access to energy services, help the environment, and boost revenues for operators. It is estimated that 485 million mobile users worldwide have no access to the electricity grid, e.g. more than 40% of Kenyans own a mobile phone, but only 23% of the population has access to the electricity grid [15]. Uganda has 30 million inhabitants, but 93% of the population has no access to electricity. They rely on small shops and kiosks that charge phones for a fee, in some cases through hook-ups to portable car batteries. This inconvenience means that most of the time the mobile phone is powered off and operators are missing revenue opportunities.

New mobile phones, chargers and charging docks attached to a base station (mobile phone tower) or a solar charging station in a village center that uses solar power can vastly expand phone usage among new, mostly low-income users of the developing world. A study conducted by the GSMA Development Fund, 'Mobile Phone Use in 2009' found strong interest in off-grid charging among mobile operators covering Africa, Asia, and Latin America. Sixty percent either already had such technologies or were investigating them. The association estimates that solar charging, if made widely available at affordable cost, could boost average revenues per phone user by 10 to 14%, and that means, off-grid charging offers a lucrative business opportunity for operators [16].

The precise value of green ICT technologies has been difficult to quantify in a simple cost-benefit analysis due to the multi-tied benefits they provide to the various sectors, the consumers, and society.

6.6 Awareness and Demand for Green Mobile Solutions

The introduction of Green ICT technologies is based on various motivations – from an awareness of environmental protection through to reducing energy costs and improving image.

From the companies' perspective, there is an increased awareness and emphasis on green initiatives, reducing the carbon footprint, and improving energy efficiency. It can be expected that with the ongoing market awareness, increasing oil prices and requirements by environmental legislation, energy efficiency will become a key aspect of companies' business strategies.

Much of the demand for Green ICT solution is coming from IT sector itself in order to increase energy efficiency of data centre technologies and electronic devices.

Other sectors of the economy will also need and demand green ICT solutions and applications to reduce environmental impacts across economic and social activities. Smart ICT solutions e.g., Smart Manufacturing, Smart Energy Management, Sustainable Energy Production, Smart Buildings & Infrastructure and Virtual Meetings will all emerge as essential for successfully meeting the demand of environmental performance, regulation and the drive to both reduce costs and maximize revenue.

An important issue in this respect is awareness of the benefits from the smart integration of ICT into new ways of operating, living, working, learning and travelling. Without the right information and incentives, it will be impossible to implement green ICT solutions in various sectors.

The consumers' demand for products with better environmental performance is, however, still very low. The low level of awareness of the benefit and opportunities of green ICT in private life is a major problem affecting consumers demand for green ICT products. On the other hand the lack of consumer demand is also a factor that can explain the delay in the roll out of innovative ICT products or the reluctance to develop eEnergy applications.

6.7 Conclusions

Above arguments for Green Mobile are presented. The need and relevance seem evident, but the impact so far has been tiny. Most of the initiatives are very recent and results may show, but real changes need a coherent effort addressing the different aspects and make them support each other. Government policies and actions can have an important influence at the national level. Considering that the global aspect is dominant in ICT in as well production and use as in environmental effects, solutions need to be found at the global level to be efficient. Global agreements and treaties are, however, likely to be quite hard to arrange at this level of details in a foreseeable future.

An obvious solution is to reveal the drivers at the sector level making the market forces – costs and revenue discussed above – drive the development. To promote this, a layered approach presented in Figure 6.2 is suggested as a framework for discussion and actions in industry fora. The approach draws on the OSI model, the categorization of different types of Green impact

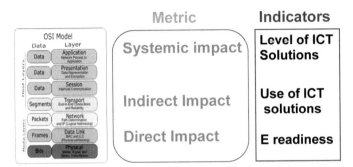

Figure 6.2 A layered approach [17].

of mobile communication combined with different indicators of the impact. Clarifying the different categories and their potentials for being mutual supportive might enhance the drive for use and development of efficient solutions green solutions based on mobile communication technologies.

References

[1] National IT and Telecom Agency, Green IT in your company – Ideas and inspiration for a greener profile, Ministry of Science Technology and Innovation, May 2009.

[2] The Climate Group, SMART 2020: Enabling the low carbon economy in the information age, 2008.

[3] ITU, The World in 2009: ICT facts and figures, Geneva, 2009.

[4] Commission Green Paper, 2000 Towards a European strategy for the security of energy supply [COM(2000)769 final], Brussels, 29 November 2000.

[5] The European Climate Change Programme, EU Action against Climate Change, European Commission, January 2006.

[6] EU Commission, Directive 2008/34/EC of the European Parliament and of the Council of 11 March 2008 amending Directive 2002/96/EC on Waste Electrical and Electronic Equipment (WEEE), as regards the implementing powers conferred on the Commission, Official Journal of the European Union, March 2008.

[7] EU Commission, Directive 2008/35/EC of the European Parliament and of the Council of 11 March 2008 amending Directive 2002/95/EC on the restriction of the use of certain hazardous substances in electrical and electronic equipment as regards the implementing powers conferred on the Commission, March 2008.

[8] EU Commission, Directive 2005/32/EC of the European Parliament and of the Council of 6 July 2005 establishing a framework for the setting of eco-design requirements for energy-using products and amending Council Directive 92/42/EEC and Directives 96/57/EC and 2000/55/EC of the European Parliament and of the Council, Official Journal of the European Union, July 2005.

[9] EU Commission, The European Commission welcomes industry's commitment to provide a common charger for mobile telephones, European Commission, Enterprise and Industry, June 2009.

[10] Milanesi, C., Gupta, A., Huy Nguyen, T., De La Vergne, H.J., Sato, A. and Zimmermann, A., Dataquest insight: Market share for mobile devices, 4Q08 and 2008, Gartner, March 2009.

[11] GSMA, Environmental Impact of Mobile Communications Devices, May 2009.

[12] Skouby, K.E., Green mobile – Is it coming? Nokia Siemens Networks, Presentation, WPMC'09, Special Session, September 2009.

[13] ABI Research, 800,000 Alternative energy-powered base stations in 2009 – Clean telecoms is the next big thing analyst insider, A weekly technology research update from ABI Research, 18 March 2009.

[14] T-Systems Enterprise Services GmbH, Green ICT: The way to green business, White Paper, Officer for Environmental Protection and Sustainability, June 2009.

[15] GSMA Development Fund, Green power for mobile: Charging Choices. Off-grid charging solutions for mobile phone, October 2009.

[16] http://www.gsmworld.com/documents/charging_choices.pdf.

[17] GSMA, GSMA research shows off-grid charging solutions for mobile phones to power US2.3bn market opportunity, October 2009.

7

ZigBee as a Key Technology for Green Communications

Christoph Spiegel[1], Sebastian Rickers[1], Peter Jung[1], Woojin Shim[2], Rami Lee[2] and Jae Hwang Yu[2]

[1]*Department of Communication Technologies, University of Duisburg-Essen, 47048 Duisburg, Germany*
[2]*SK Telecom, Convergence & Internet R&D Center, Seoul 100-999, Korea*

Abstract

The IEEE standard 802.15.4 for low-rate wireless personal area networks and ZigBee will play an important role for low power networks. From the authors' point of view, ZigBee might even become a key technology for such networks. Since there are numerous "green" feature sets defined, ZigBee can also be regarded as an enabler for Green Communications. A low complexity and low power implementation of a ZigBee receiver which is driven by zero-crossing demodulation (ZXD), an irregular sampling technique, will be introduced in this chapter.

Keywords: IEEE 802.15.4, ZigBee, zero-crossing demodulation (ZXD), system on a chip (SoC), sensor networks, low power networks.

7.1 Introduction

In the last decades, there has been a growing demand to rethink the relationship between mankind and the environment. The emerging need for energy contradicts the awareness that only sustainable exploitation of our planet's resources will have the potential to advert or at least damp the global warm-

R. Prasad et al. (Eds.), Towards Green ICT, 87–96.

ing and its late effects. In this context, besides the accessibility of renewable energy, novel technologies are required to drastically increase the efficiency of energy utilization.

Smart grids are commonly regarded to play a key role for streamlining our today's energy grids. Besides the power transmission and distribution grids, also the in-house grids have to be prepared before they can become part of large-scale smart grids. To facilitate this requirement, ZigBee can serve as a key technology enabling networking between the different devices connected to the power grid. ZigBee was developed for wireless devices requiring low data rate communications in conjunction with ultra low power consumption. Following the official statements of the ZigBee Alliance [1], ZigBee devices shall be targeting the markets consumer electronics, energy management and efficiency, health care, home automation, building automation and industrial automation. This shows that from the very beginning of ZigBee, Green Communications have been considered as a potential market.

This chapter is structured as follows. Section 7.2 gives a general description of the IEEE 802.15.4 standard and ZigBee. Section 7.3 explains how irregular sampling, especially zero-crossing demodulation (ZXD) can help to establish lowest cost, lowest complexity receivers capable in receiving IEEE 802.15.4 compliant signals. In Section 7.4, an integrated circuit operating with ZXD is introduced. Section 7.5 concludes the previous sections and Section 7.6 finally gives an outlook for the future of ZigBee in the scope of Green Communications.

7.2 The IEEE 802.15.4 Specification and ZigBee

The IEEE 802.15.4 specification extends the range of WPAN (wireless personal area network) standards (also including IEEE 802.15.1 which is derived from the Bluetooth specification) by a low data rate system whose design criteria comprise low complexity and a high degree of robustness. IEEE 802.15.4 defines multiple physical layers (PHYs) using the modulation schemes offset quaternary phase shift keying (OQPSK), binary phase shift keying (BPSK) and amplitude shift keying (ASK), each featuring a modulation order of one single bit per symbol. The supported data rates range from 20 up to 250 kbit/s. The supported frequency bands cover the world-wide 2450 MHz ISM (Industrial, Scientific and Medical) band and the regional 868 and 915 MHz bands.

However, not all PHYs support all bands in all countries and there are also optional PHYs which do not necessarily have to be implemented in devices to

Table 7.1 Data rates for common wireless systems.

System	Maximum data rate
WiFi (802.11n)	600 Mbit/s
WiFi (802.11g)	54 Mbit/s
WiFi (802.11b)	11 Mbit/s
Bluetooth 2.0 EDR	3 Mbit/s
ZigBee (802.15.4)	250 kbit/s

obtain IEEE 802.15.4 compliance. However, the OQPSK modulation scheme in the 2450 MHz band with a symbol rate of $f_s = 1/T_s = 2$ MHz is the most popular mode and is therefore focused on in the remainder of this chapter. It will be termed 'native mode' in what follows and explained in more detail in the following paragraphs.

When comparing IEEE 802.15.4 to other popular WLAN (Wireless Local Area Network) and WPAN standards such as WiFi and Bluetooth (Table 7.1), respectively, it becomes apparent that the maximum data rate defined in IEEE 802.15.4 is well below any other state of the art wireless system. This is the price to pay to reach a high degree of robustness and simplicity.

In order to obtain this high degree of robustness, the ZigBee native mode uses a spreading technique similar to direct sequence spread spectrum (DSSS) where groups of four information bits b_k are mapped to 1 out of 16 orthogonal spreading sequences with 32 chips length each. This corresponds to a spreading factor $SF = 8$ yielding a spreading gain of 9 dB in the logarithmic domain.

After spreading the information bits, OQPSK modulation using a half-sine pulse shaping filter is applied to the resulting chip sequence. The pulse shaping filter is of length $2T_s$ and is specified as

$$h_s(t) = \sin\left(\frac{t\pi}{2T_s}\right) * \text{rect}\left(\frac{t - T_s}{2T_s}\right). \tag{7.1}$$

When using OQPSK, the information chips are mapped to the in-phase and quadrature component in an alternating fashion. The complex-valued OQPSK baseband transmit signal $\underline{s}(t)$ is defined in (7.2) where $c_k \in \{+1, -1\}$, $0 \le k \le K - 1$ is the length K information chip sequence and $T_s = 1/f_s$ is the symbol period. In (7.2), $\delta(t)$ denotes Dirac's delta distribution and $h_s(t)$ is the pulse shaping filter of (7.1). The convolution operation is represented by $*$, mod is the integer modulus operation and $\lfloor \cdot \rfloor$ returns the integer floor of its argument.

$$\underline{s}(t) = \sum_{k=0}^{K-1}(-1)^{\mathrm{mod}(\lfloor k/2 \rfloor, 2)} \mathrm{j}^k \delta(t - kT_{\mathrm{s}})c_k * h_{\mathrm{s}}(t). \tag{7.2}$$

Equation (7.2) yields a modulation scheme which is identical with minimum shift keying (MSK) which is known to be a continuous phase modulation (CPM) scheme. Therefore, (7.2) can be rewritten as

$$\underline{s}(t) = \exp(\mathrm{j}\varphi(t)), \tag{7.3}$$

where

$$\varphi(t) = \int_0^t \sum_k^{K-1} c_k \frac{\pi}{2T_{\mathrm{s}}} \mathrm{rect}\left(\frac{t - (k + 1/2)T_{\mathrm{s}}}{T_{\mathrm{s}}}\right) \tag{7.4}$$

is the continuous phase of the transmit signal showing a fixed phase deviation of either $+\pi/2$ or $-\pi/2$ per symbol period T_{s}.

In order to obtain the RF (radio frequency) domain transmit signal $s_0(t)$, the complex baseband signal of (7.3) has to be upconverted to the desired RF carrier frequency f_0:

$$\begin{aligned} s_0(t) &= \Re\{\underline{s}(t)\exp(\mathrm{j}2\pi f_0)\} \\ &= \Re\{\underline{s}(t)\}\cos(2\pi f_0) - \Im\{\underline{s}(t)\}\sin(2\pi f_0). \end{aligned} \tag{7.5}$$

Finally, (7.5) can also be regarded as continuous phase frequency shift keying (CPFSK) [2] which is a special kind of two-state frequency shift keying (2FSK) with a frequency deviation of $\Delta f = f_{\mathrm{s}}/4$. This property enables the design of simple transmitter as well as receiver structures. The transmitters can be implemented as digital frequency synthesizers, the receivers as frequency discriminators [3]. In a nutshell, the OQPSK modulation scheme specified in IEEE 802.15.4 can be regarded as MSK and thus enables low complexity and low cost wireless devices [4].

IEEE 802.15.4 only defines the PHY layer and the MAC (medium access control) sublayer, i.e. the part which handles the access of multiple devices of the same network to the shared medium. In contrast, ZigBee defines the layers above these layers, i.e. the layers which rule the behavior of the different device types specified by the ZigBee Alliance [1]. There are certain device profiles defined, such as the smart energy profile and the home automation profile. The ZigBee part of the overall protocol stack features the network, the security and the application layer.

7.3 ZigBee and Zero-Crossing Demodulation

ZigBee devices are required to consume ultra-low power. Medical monitoring devices can be taken as an example: They are mostly equipped with a battery which is expected to last up to ten years or even more. This requirement can only be fulfilled if the wireless modem has a very low complexity. At this point, ZXD comes into play and shall therefore be explained briefly in what follows.

Conventional receivers operate with regular sampling. The classical approach is to downconvert the receive signal from the RF domain to the baseband and then sample both in-phase and quadrature component with a regular, i.e. an equidistant sampling grid. This procedure requires two common analog to digital converters (ADCs) with a certain resolution, i.e. with a certain number of output bits. For such regularly sampled systems, digital receiver algorithms including synchronization, channel estimation and data detection are well known and can be considered state of the art.

In contrast to regular sampling, irregular sampling techniques are defined as schemes not using a regular sampling grid but instead any other kind of sampling grid. For the special case of ZXD, the irregular sampling grid is determined by the receive signal itself. More precisely, the sampling instants are determined by the positions where the receive signal amplitude crosses the zero threshold. It was shown in previous analyses that either rising or falling transitions of the receive signal should be taken into account and that the receive signal must be located at an appropriate low intermediate frequency (IF) to make it suitable for ZXD [5]. In order to obtain the IF domain signal $s_{IF}(t)$, the RF domain signal of (7.5) must be down-converted yielding the desired low IF signal centered around f_{IF}.

The advantage of ZXD receivers compared to conventional ones is the reduced complexity of both the radio front-end as well as the digital part of the receiver. The radio front-end can be designed as a low-IF receiver requiring only one signal path and thus being free of deficiencies such as IQ imbalance and DC offset. Moreover, the two multi-bit ADCs required for the regular receiver can be replaced by a simple comparator making the analog front-end less expensive.

Figure 7.1 illustrates how to obtain digital data for further processing from the analog IF domain receive signal $s_{IF}(t)$. First of all, there is a counter running in endless cycles at a clock frequency $f_{clk} = 1/T_{clk}$. How to choose this design parameter appropriately was already shown in previous publications along with important notes on real-time implementations of ZXD [6].

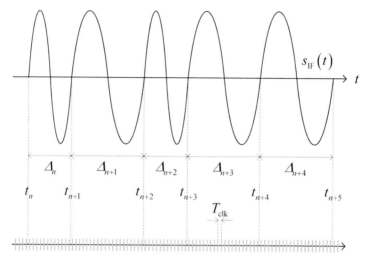

Figure 7.1 Illustration of zero-crossing demodulation.

Whenever a rising-edge zero-crossing of the low IF receive signal occurs (time instants t_n in Figure 7.1), the current counter value is sampled into a buffer and can be forwarded to the ZXD demodulation unit for further processing.

The simplest method allowing the recovery of the transmitted chips is frequency discrimination. This can easily be accomplished by means of a device detecting whether the current symbol has a frequency higher or lower than the chosen intermediate frequency (cf. Section 7.2) which is the average frequency at the same time. Since the differences between consecutive zero-crossings (Δ_n in Figure 7.1) can be regarded as a measure for the reciprocal instantaneous frequency of the receive signal, the frequency discriminator can be implemented at a very low complexity in a purely digital fashion [3].

It shall be noted that ZXD is an incoherent demodulation approach in principle making it inferior to classical coherent demodulation schemes. However, also for ZXD there are more sophisticated demodulation schemes which yield a better performance than the scheme described before, especially in realistic environments without line of sight and with multi-user effects like adjacent channel interference. They have been described extensively in the authors' earlier works [7–10].

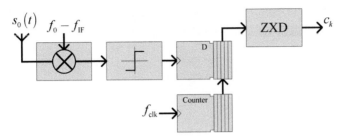

Figure 7.2 Block diagram of the ZXD receiver.

7.4 Design of an Integrated Circuit

In order to obtain the best set of design parameters with respect to the IF at which the signal is presented to the receiver and the selection of appropriate system clocks, numerous simulations are required. These simulations have been carried out by means of the authors' ZXD simulation framework written in pure ISOC language. Once the optimum set of parameters is determined, the design of an integrated circuit can be started.

Figure 7.2 shows the simplified block diagram of the ZXD based receiver comprising an analog downconversion unit, a comparator, a cyclic up-counter, a Dtype latch and the actual ZXD demodulation block. The downconversion unit mixes the receive signal $s_0(t)$ from the RF domain (center frequency f_0) down to the desired IF domain (center frequency f_{IF}). The resulting signal is taken as input for the comparator. The comparator's output signal triggers the Dtype latch on each rising edge. The latch's data input is fed by the output of the cyclic up-counter clocked at f_{clk}. The buffered counter word is then presented to the ZXD block which carries out frame and chip synchronization, frequency error correction and recovery of the transmitted chips c_k. The chips can then be despreaded (cf. Section 7.2) to finally retrieve the transmitted information bits sequence.

The developed chip is a complete system on a chip (SoC) featuring a full ZigBee compliant transceiver including the radio front-end and the IEEE 802.15.4 PHY and MAC layer. A photograph of a test board with a chip sample assembled is shown in Figure 7.3. Besides the SMA connector used for RF signal input, it is equipped with only very few auxiliary components. Since the chip is highly integrated, the RF signal can be almost directly fed into the device. It features a standard serial bit interface to directly communicate with attached data sources or data sinks, depending on whether the chip is used in transmit or receive mode.

Figure 7.3   Test module with a sample of the ZXD based ZigBee SoC assembled.

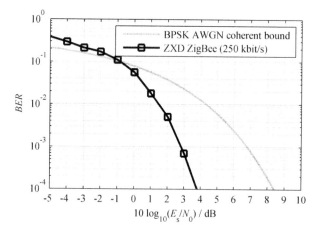

Figure 7.4   Performance results of the ZXD ZigBee receiver.

The digital part of the receiver was designed using the hardware description language Verilog. Based on the design, simulation results for the receiver
have been obtained considering all implementation effects such as the fixed-
point representation of the signals, limited clock rates and timing effects. The
resulting performance is shown in Figure 7.4.

Compared to the AWGN (additive white Gaussian noise) bound of a
coherent BPSK (binary phase shift keying) receiver, the authors' ZXD implementation shows a gain of 4.5 dB at a reference bit error ratio (BER) of
10^{-4}. With respect to the expected spreading gain of 9 dB (cf. Section 7.2),
the overall loss turns out to be 4.5 dB. The largest contribution comes from
the circumstance that ZXD is an incoherent demodulation technique (cf. Section 7.3), being approximately 3 dB worse than an optimum coherent receiver.
The remaining portion of 1.5 dB can be regarded as implementation loss.

7.5 Conclusion

The authors showed that ZXD based receivers are suitable for the use in Zig-Bee devices. ZXD based receivers are more cost-effective than conventional receivers at the expense of slightly reduced performance compared to regular sampling receivers. This is because ZXD receivers are incoherent in principle.

Besides the various simulations the authors carried out to show the feasibility of their concept and to obtain appropriate design parameters, an integrated circuit (ZigBee SoC) was built featuring a ZXD based receiver. The chip was recently verified and shows the expected performance proving the previously established ZXD concept also for ZigBee compliant receivers.

7.6 Outlook

Recently, the ZigBee Alliance released the new green power feature set [11] enabling the supply of ZigBee devices only by energy harvesting techniques. These techniques include the utilization of ambient energy such as temperature and vibrations making the presence of a battery in each device superfluous.

This development strengthens the authors' belief that especially in the field of Green Communications, ZigBee will play an important role. Hence it will be required to further optimize the energy consumption of future ZigBee transceivers and maximize their overall energy efficiency to make them functioning even without a constant power supply.

The authors are convinced that irregular sampling will be one possible key technology paving the way towards such devices. However, further investigation on the topic of energy efficiency is still required and will be tackled in the near future.

Acknowledgements

The findings presented in this chapter arise from a joint project of the Department of Communication Technologies (Lehrstuhl für Kommunikationstechnik) at the University of Duisburg-Essen, Duisburg, Germany with the AT Development Group at SK Telecom, Seoul, Korea. The Department of Communication Technologies is very grateful for the strong support they gain from SK Telecom and would like to thank SK Telecom for their continuous collaboration.

References

[1] ZigBee Alliance, Homepage, `http://www.zigbee.org`.

[2] Proakis, J.G., *Digital Communications*, Fourth Edition, McGraw-Hill, 2001.

[3] Garodnick, J., Greco, J. and Schilling, D.L., Theory of operation and design of an all digital FM discriminator. *IEEE Transactions on Communications*, **20**, 1159–1165, 1972.

[4] Scholand, T. and Jung, P., Advanced intermediate frequency zero-crossing detection of bandpass filtered MSK signals. *Electronics Letters* **39**, 736–738, 2003.

[5] Waadt, A., Scholand, T., Spiegel, C., Burnic, A. and Jung, P., Optimal choice of the intermediate frequency for zero-crossing detectors. *Electronics Letters* **43**, 678–680, 2007.

[6] Scholand, T., Spiegel, C., Waadt, A., Burnic, A. and Jung, P., A real-time zero-crossing demodulation concept. *Journal of Wireless Personal Communications* **43**, 157–183, 2007.

[7] Scholand, T. and Jung, P., Novel receiver structure for Bluetooth based on modified zero-crossing demodulation. In *Proceedings of the IEEE Globecom 2003*, San Francisco, USA, 2003.

[8] Scholand, T., Waadt, A. and Jung, P., Maxlog-ML symbol estimation postprocessor for intermediate frequency LDI detectors. *Electronics Letters* **40**, 183–185, 2004.

[9] Scholand, T., Faber, T. and Jung, P., Novel Bluetooth receiver structure deploying zero-crossing demodulation with zero-forcing equalization. In *Proceedings of the International Conference on Computers and Devices for Communication (CODEC 2004)*, Kolkata, India, 13 January 2004.

[10] Scholand, T., Spiegel, C., Waadt, A., Burnic, A. and Jung, P., Applying zero-crossing demodulation to the Bluetooth enhanced data rate mode. In *The 17th Annual IEEE International Symposium on Personal, Indoor and Mobile Radio Communications (PIMRC2006)*, Helsinki, Finland, 11–14 September 2006.

[11] ZigBee Alliance, New ZigBee green power feature set revealed. Press Release, 29 June 2009.

8

Using ICT in Greening: The Role of RFID

Rasmus Krigslund[1], Petar Popovski[1], Iskra Dukovska-Popovska[2],
Gert F. Pedersen[1] and Boris Manev[3]

[1]*Department of Antennas, Propagation and Radio Networking, Aalborg University, 9220 Aalborg, Denmark*
[2]*Department of Production, Aalborg University, 9220 Aalborg, Denmark*
[3]*Institute for Environmental Studies, Free University Amsterdam, 1081 HV Amsterdam, The Netherlands*

Abstract

In recent years, the awareness about the harmful long-term effects of the greenhouse gases (GHG) has markedly increased. As information and communication technologies (ICT) have successfully and irreversibly pervaded everyday life, many technology leaders have started to look into the green ICT, i.e. ICT solutions that have reduced carbon emissions. A sobering report from WWF has indicated that only 2% of the carbon emissions can be attributed to the ICT systems. Therefore, the real green impact of ICT, at least in short term, can be made by using ICT to make the other systems more energy-efficient, such as the buildings, transportation, power grid, industry and production. In this chapter we argue that the RFID technology can be one of the key drivers for implementing green ICT solutions in the systems that are not originally associated with ICT, e.g. transportation. In particular, the passive, battery-less RFID systems have potential to map the physical world to the virtual one and drive actions back from the virtual to the physical world with positive carbon balance, i.e. the carbon emissions saved by those actions are larger than the carbon emissions caused to operate the RFID systems. Motivated by that observation, we list several example green applications of the RFID systems.

R. Prasad et al. (Eds.), Towards Green ICT, 97–116.

Keywords: Green ICT, RFID, energy harvesting, wireless sensor networks.

8.1 Introduction

The information and communication technologies (ICT) have brought profound improvements in all the areas of human activity and the quality of life. The latter relies mostly on impact that the ICT has had on the people through the immediate rewards provided by those systems. For example, the mobile phone brings reachability, which is almost ubiquitous, while the Internet provided unprecedented access to information and created countless opportunities for novel social interactions. Nevertheless, ICT has yet to unleash its potential in solving the global problems that will have long-term effects, such as the reversal of the trends in global warming, reduction of greenhouse gasses (GHG), dealing with the scarce resources (energy, water), protection of the environment, and sustainable global growth. These objectives are significantly interdependent, while the reduction of carbon emissions will have the largest effects.

Carbon dioxide (CO_2) is the most important anthropogenic greenhouse gas (GHG) accounting for almost 80% of the overall GHG emissions in the atmosphere [6]. The usage of ICT for reducing carbon emissions has been a subject of many recent analyses, see [24] and the references therein. Greening is also seen as one of the impulses that can direct the research and innovation within ICT in the coming years. The understanding of the interaction between ICT and GHG is still in its infancy, although some indicators are already emerging. For example, the WWF report [24] has indicated that only 2% of the current carbon emissions can be attributed to the ICT systems. This is a strong message that, in a short term, the benefits obtained from making, for example "green wireless networks" or "green server farms", can be almost negligible as compared to the benefits obtained if the ICT actions are targeted to ameliorate the activities that are responsible for the dominant part of the carbon emissions, such as the power production sector, buildings, industry and production, transport, etc. On the other hand, the volume of devices related to ICT will increase rapidly in the coming years, and, if those computing/communication devices remain to operate with the same power budget as today, then it is clear that the carbon footprint of ICT will be much larger than 2%. In that sense, one important research areas remains to be how to make ICT green and more energy efficient, in order to be prepared to sustained the incoming proliferation of computing/communication devices.

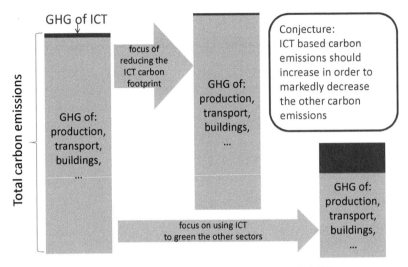

Figure 8.1 The RFID tag and reader comprising a simple RFID system.

In this chapter we focus on the problem how to use ICT for greening rather than greening ICT itself. The overall task of putting ICT to work towards making the world greener is enormously complex and includes several subtasks. The first is data collection and analysis, in order to make assessment of the GHG production in relation to all the human activities, and thus make projections about the quantitative impact that the ICT will have on the GHG emissions rooted at those activities. Another subtask is development of the technology that will enable low-carbon operation in the targeted activity areas, as well as innovation of applications that can replace some of the present activities with low-carbon counterparts. The "usual suspect" related to the latter is the replacement of the travels by video-conferencing, but many more applications and examples are needed to drive the environmental impact.

A proper milieu for green ICTs to be put to work can be created by suitable policies and strategies. Such policies should be created by accounting for the opportunities and capabilities of the technology. For example, the system of incentives that can be introduced regarding the GHG activities can be built around rules for monitoring/acting that are assuming certain level of technology sophistication. Using ICT to reduce the overall emissions has a great potential, as illustrated in Figure 8.1.

In this chapter we advocate the Radio Frequency Identification (RFID) technology as one of the key components of the green ICT. Before summarizing the arguments behind such a claim, we need to briefly introduce the RFID technology. A RFID system provides automated identification and information gathering from objects and people. The two key components of an RFID system are tags and readers. A tag is a small microchip equipped with antenna which is attached to the physical object or the person. The most interesting types of tags are the passive, batteryless tags, as they can have low cost and thus be deployed in large volumes. In fact, batteryless, passive devices capable of computing and communication are by definition compatible with green requirements. The readers (or interrogators) are devices, usually deployed at strategic locations in order to efficiently collect information from the tags in their radio range. The tags are attached to the objects or humans and thus make the physical world perceptible for the computers. That is why the passive tags are enablers of the "Internet of things". Particularly important developments related to the RFID technology can be seen in RFID sensors and energy harvesting. With RFID sensors, the tag is integrated with a sensor or multiple sensors and the radio link to the reader is used to convey the sensed data. Furthermore, energy harvesting will augment the computation/communication capabilities of the passive tags.

We argue that the role of RFID in enabling green ICT is multilateral. In the data collection process, it enables gathering of information to a very detailed level, which is an important step to relate different activities to the GHG. The usage of RFID tags can have direct impact on the environment by enabling more efficient waste management and recycling. Being the bond between the physical and the cyber world, the tags can be the key drivers in attacking the 98% opportunity of the carbon emissions, by enabling real time planning in the virtual domain and sensing/acting in the physical domain. Finally, the passive tags can enable enforcement of policies and strategies and be instrumental to introduce incentive systems in GHG emission and the carbon trading schemes.

Before describing the details of how RFID has the potential to reduce carbon emissions, in the next section we give a brief overview of the current global situation with respect to GHG emissions.

8.2 Carbon Emissions: Fact and Figures

Carbon dioxide (CO_2) is the most important anthropogenic greenhouse gas (GHG) accounting for almost 80% of the overall GHG emissions in the

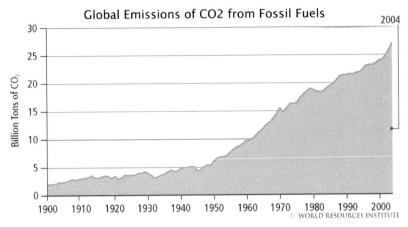

Figure 8.2 Historical development of the carbon emission volumes. From: http://cait.wri.org/figures.php?page=ntn/1-2.

atmosphere. According to statistics by the International Panel on Climate Change (IPCC) the global atmospheric concentration of CO_2 has increased to 379 ppm in 2005, which exceeds by far the natural range over the last 650,000 years (180 to 300 ppm). The concentration of CO_2 has grown with increased intensity during the last decade. Thus, while from 1960 to 2005 CO2 concentration had an average growth rate of 1.4 ppm per year, only from 1995 to 2005 the average growth rate has been 1.9 ppm per year [22]. This is a direct consequence of the fact that annual fossil CO2 emissions increased from an average of 23.5 billion tons per year in the 1990s, to 26.4 billion tons of CO_2 per year in 2000–2005, see Figure 8.2.

8.2.1 Sector Approach

Statistical data shows that largest CO_2 emitter is the power production sector with 28% of total emissions [6]. Another 32% of the global CO_2 emissions coming from industry, transport, and residential services and agriculture, are emitted by power consumption when energy is processed into final products. Nevertheless, attention should not be concentrated only on CO_2 emission but also on other GHG gasses such as CH_4, N_2O, SF_6, HFC, and PFC as they account for almost one fourth, or 23% of the total GHG emissions. Sector-emission percentages are as listed in Table 8.1.

Table 8.1   Sectoral emission in percentage [6].

Power production	28%
Industry	12%
Transport	13%
Residential services and agriculture	7%
Land use change	17%
Non-CO_2 GHG emissions	23%

8.2.2 Intervention

By 2050 global emissions of CO_2 need to be reduced by 85% in order to keep the temperature increase below 2°C. To achieve this, interventions in each of the GHG emitting sectors is needed by improvement of energy efficiency; increase in the share of renewables in the energy-mix; as well as "cleaning" of the fossil-fuel energy generation sector. Increase in efficiency is mostly relevant in the sectors such as industry, buildings, transport and power generation where energy and materials are being transformed into products and services [6]. The power production sector has on average very low efficiency. Old coal power plants have an average efficiency of 30%. Yet, with new components and optimized process integration, their efficiency could improve up to 50%, which would mean a reduction of 1.4 million tons of CO_2 annually [6].

Electric grids, also, contribute to the loss of efficiency. During the transportation of electricity to consumers certain percentage is lost due to electrical resistance in the cables in the grid. Future development of Smart Grid power network will provide sustainable development, efficiency and cost benefits [14]. The Smart Grid leverages on ICT concepts in order to provide information flow and control over the electric grid in order to make the grid more efficient.

A very important and promising sector, where energy efficiency can be increased while CO_2 emissions can be significantly reduced is the building sector which is accountable for one third of the global energy related CO_2 emissions [33]. Energy efficiency in both commercial and residential buildings can be improved by using variety of technologies, designs and materials. Some technologies and concepts for improvement of energy efficiency are through efficient lightning, for example Light-Emitting Diode (LED). Other interventions that can contribute to energy efficiency include introduction of equipment with lowest stand-by power, reduction of energy outside office hours, efficient cooling/heating. A good example for such an energy efficient building is the passive house which can save up to 75% energy consumption and thus decrease its carbon footprint [13].

Improvement in energy efficiency requires investments in the corresponding R&D activities. Since 20% of the world's population that is mostly located in the developed countries consumes 80% of the world's natural resources including energy, they should be accountable for investment in energy efficiency. Yet, according to predictions by the US Department of Energy, world's energy consumption will grow by 50% from 2005 to 2050 [3], mostly due to increased demand in emerging economies such as China and India. Developed countries will have to commit to invest in improvement of their energy efficiency, yet will also have to invest in improvement in the energy efficiency in the developing world without affecting their path to development. According to Greenpeace, G8 countries will only put the Post-Kyoto negotiation on a path towards success if they commit to contribute USD 106 billion every year by 2020 to help developing countries face the climate change challenges [16].

The dichotomy related to the issue in which countries the investment in energy efficient technologies should be dominant, can be mapped into a dichotomy of two radically different technology approaches. In the developing countries many of the systems (which are major sources of carbon emissions in the developed countries) have not yet been deployed. This gives opportunity for a *clean slate system design*, which starts to build the power system, transportation, industry, etc. based on the level of technological sophistication that is available today or in the near future and, most importantly, not being strongly bound by backward compatibility. Such a clean slate design can, for example, result in a transportation system that is as flexible and personalized as the car, but far more friendly to the environment. In such newly developed systems, ICT will have a decisive role, as the systems will rely on rich information flow and control among the system components.

Contrary to the clean slate approach, in the developed countries there are systems that are widely deployed and cannot be completely and instantaneously replaced, but they need to be gradually evolved, accounting for the backward compatibility. Hence in this case we need *wean slate system design*, where the existing systems should be upgraded to be weaned on the requirements for low carbon emissions.

These two approaches have different perspective, but the same goal, i.e. reducing emissions. With this overview of the current situation with respect to emissions we have identified in which sectors the largest reduction can be

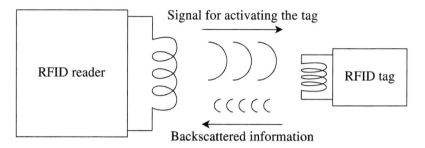

Figure 8.3 The RFID tag and reader comprising a simple RFID system.

achieved. In the following we present ICT technologies that have the potential of a great green impact on these sectors.

8.3 RFID and Some Related Technologies

There exist several opportunities for using ICT to reduce carbon emissions. Some of these are listed in [24]. One example is to use ICT to optimize scheduling and resource allocation in the physical world. In order to do this the physical world must be made perceptible in the digital world. As an example consider Smart Grid, where measuring devices are sampling the current power usage of a building and maybe estimates near-future power requirements, e.g. by knowing which devices that are currently using power [27]. Using communication systems this information is fed back to the supplier, who can then determine how to distribute the power to the customers most efficiently. A promising technology for coupling the physical world with the digital world is Radio Frequency IDentification (RFID).

8.3.1 Basics of RFID Systems

RFID technology has been around for quite some time now, but recently this area has received immense attention due to reduced costs of implementation and production. An RFID system has two basic parts: Readers and Tags, as illustrated in Figure 8.3. The reader is a transceiver that can activate the tag and reads its content. A tag is a device small enough to be embedded into an object, and it consists basically of a microchip with modest storage capacity connected to an antenna. When activated on demand by an external reader the tag backscatters its unique identifier and the information saved in its memory.

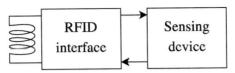

Figure 8.4 The blocks of an RFID sensor device.

There exist different types of tags: Active tags, where an internal power source allows it to transmit its information at any time. However, having its own power source puts a limit to the lifetime of the tag, but it increases the communication range of the tag. The tags that have large potential for wide deployment and usage are the passive tags, which are powered by inductively coupled power from the signal transmitted by the reader. In other words, the passive tags harvest the required energy from the over-the-air transmission by the reader. By omitting the internal power source decreases the production cost and increases the lifetime of the tag.

A reader does not require line-of-sight in order to identify a tag and read its content which means that RFID is a technology which can be utilised for many different purposes, and in many different scenarios. If physical objects are tagged and readers are placed in suitable locations the presence of unique objects can be identified along with the information stored in their memory. Moreover using RFID sensors and actuators it is not only possible to identify the presence of physical objects, but also interact with the physical world.

8.3.2 RFID Sensor Systems

Wireless sensor networks consist of a large number of small sensing, self-powered nodes that gather information and ultimately transmit this information to a base station in a wireless fashion. [26] Recently battery free wireless sensors based on RFID has been considered. An RFID sensor is basically a passive RFID tag paired with a small sensing device. In this way the resulting device is a uniquely identifiable sensor with a wireless interface, as illustrated by the block diagram in Figure 8.4. An example of such a device has been presented in [28].

This means that the advantages of RFID is brought to wireless sensor networks creating devices with a small form factor and a long lifetime. Similar to the RFID system described in Section 8.3.1, the RFID sensor is powered from inductive coupling when it is in the range of a transmitting reader, and the RFID sensor then backscatters its unique identifier and the state of the sensor. Since readers are not transmitting continuously, the power source for an RFID sensor is intermittent and unpredictable. This type of communication provides sufficient power for standard RFID systems, where a tag only backscatters the identifier. However, RFID sensors requires more power to operate the sensing device compared to standard tags. Hence it is difficult for the RFID sensor to assure that its tasks are completed based on the power received from the reader. It can therefore be beneficial to consider harvesting energy from other sources than the transmitted power from the reader [7].

8.3.3 Energy Harvesting

Recently energy harvesting has been considered in order to prolong battery life of mobile phones [15], but in the context of small wireless devices, e.g. RFID sensors, it has the potential to make batteries in small wireless devices obsolete. The concept of energy harvesting can therefore decrease production costs as well as the required level of maintenance, while increasing the lifetime of the wireless sensor networks.

There exist several options for harvesting energy from the surroundings [25]. For example energy from light, which requires a large surface, or vibrations which requires mechanical apparatus. However, for wireless sensors a small form factor is desired, hence light and vibrations are not suited as energy sources for this application. Instead, inductive coupling may be an interesting option. In today's RFID system the tag uses such power to backscatter its identifier to the reader, but this might evolve towards tags that store such energy for later usage. Considering the ubiquity and abundance of various wireless transmitters, a wireless device is not confined to harvest energy from the communication with its own base station, but has the possibility of continuously harvesting energy from ambient transmitters.

8.4 Green Applications Involving Passive Wireless Devices

We have already stated that the potential of RFID regarding green operation of various systems is seen in the fact that RFID provides the link between the physical and the virtual world. In this section we will concretize this, rather abstract, claim through three example green applications of the RFID system or, more general, systems with passive wireless devices.

8.4.1 Supply Chain Management

RFID has been initially used in the logistics operations of the supply chains to track and manage assets through railroad, shipping and trucking. Due to the emergence of passive tags and technology costs decrease, the use of RFID has been expanded through the different parts of the supply chain. Some of the frequently reported RFID enabled improvements have been: automation of inventory update and discrepancies detection, liberation of considerable human labor from certain workflows, as well as decrease of the possibilities of human errors in repetitive activities. Another significant advantage of RFID is the potential for objects and associated process/environment information visibility to all the participants through the supply chain.

Nevertheless, there are quite few RFID applications that enable the greening aspect of the supply chains. *Reverse logistics*, i.e. from customer to manufacturer or waste centres, has been traditionally seen as the "green" part of the supply chain. Therefore, most of the reported RFID applications in greening the supply chain are related to reverse logistics. The advantages of implementing an RFID-based system are through the increase of recovered products, tracking of returned products, simplification of the operations of collecting, sorting, and disassembly, as well as reduction of the quantities of toxic components in the environment. Overall, the RFID enables minimization of the complexities of the reverse logistics processes that comes mainly from irregular material flow as well as inventories that are in different conditions, e.g. repaired, defective, damaged, etc. RFID combined with other computational intelligence techniques [21] can help determine the economical transportation from collection points to collection centers. It minimizes the holding time and depreciating value for the returned products at the same time. Therefore, an improved reverse information flow supported by RFID would reduce resource consumption, e.g. transportation, storage, obsoleteness, and thus reduce the environmental impact in the total product lifecycle. However, there are no case applications that report emissions and resource reductions enabled by RFID.

Greening related applications of RFID in forward supply chain, i.e. from producer to customer, have not been reported up to our knowledge. Nevertheless, some potentials of the technology can be speculated. RFID technology can be utilized as a diagnostic mechanism by enabling screening of different processes through the supply chain over any period of time (days or weeks) in order to identify the emission rates and energy efficiency of different components of the process. This information could be used in re-planning and re-designing of different processes and the supply chain as a whole with the aim of reducing resource consumption and GHG emissions.

In logistics, a trade-off between warehouse size and increased transportation needs to be evaluated, and RFID can help in diagnosing different situations. Although the general logic has been towards decreasing stocks at different points, such philosophy leads to increased inventories in transport, which is affecting the environment negatively. RFID can enable significant reductions in logistic-related costs by helping eliminating unnecessary transportation and finding the optimal mode of transportation for all shipping. This can provide significant carbon benefit and can be perceived as being green, and thus contribute to the price of the product or the carbon tax of the product. Clearly, such carbon-monitoring RFID records should be readable only by authorities. It is easy to see that such a monitoring can be a strong incentive for all involved parties to introduce low-carbon practices for each product.

Regarding degree of utilization of transport equipment, a tag can be envisaged which carries the information about the dimensions, weight, or volume of the package. A tag could also be placed at the container carrying the information about the volume/dimensions of the container. In such way, an explicit contribution is created from the use of RFID technology, by enabling higher utilization of the transport equipment. Moreover, with respect to food products complete control over the supply chain and production line will help reduce waste. The most significant environmental effects are not achieved by treating the end-of-pipeline waste, but rather the inherent upstream wastes related to cultivation of crops and livestock, processing of these, production of additives etc. When food is wasted, all activities upstream and their related emissions are in vain [23]. Therefore, an RFID supported systems can enable dynamic planning and fulfilment of the demand, which would decrease the wastes across the whole chain. Another application is envisaged in the case of accidents in the food supply chains. As an example consider the case where a number of cartons with milk are damaged due to detergent in the milk. If RFID is utilized it is possible to uniquely identify and destroy only the inflicted cartons which reduce waste in the production line. This does not have a direct impact on carbon emission, but it reduces the average carbon footprint for each successfully produced object and it provides more security for the consumers. Furthermore, it is green in a sense that it has environmental impact.

8.4.2 Smart Buildings

Implementing RFID technology in buildings and combine the information gathered by the RFID readers with the installed appliances can realize what is referred to as Smart Buildings. As an example consider an office building equipped with an RFID sensor network. In addition each employee is equipped with an RFID tag, e. g. in their name tag or access card. Using these tags the sensor network installed in the building can adjust the utilized resources in order to minimize carbon emission. As an example the air condition could be controlled by temperature sensors and the light installations could sense if any tagged employees were in its proximity. If not, the light should be dimmed, or completely turned off.

Current implementations of wireless sensor networks in buildings show a decrease of approximately 20% in energy consumption [31]. Comprehensive control of utilized resources in buildings therefore possess a large potential when a reduction in carbon emissions is desired. Moreover, in addition to reduced energy consumption, wireless sensor networks also enable low cost monitoring of occupants, e. g. elder citizens living at home, in order to alert care givers in the event of accidents or illness [10], which indirectly has green effect, by sparing the transportation for regular personal visits.

8.4.3 Intelligent Transportation

Sensor networks can be used to make transportation intelligent in order to reduce carbon emission. For example a sensor network distributed throughout the infrastructure makes it possible to gather information on for example roadwork and traffic load. This data can then be used by the navigation systems in each vehicle to plan the most energy efficient route, e. g. by avoiding traffic jams. Moreover, implementing systems for communication between cars and traffic lights makes it possible to automatically turn off the engine in the cars waiting at a red light. This will reduce the time engines are running idle, which reduces carbon emission. This would also enable traffic lights to sense the traffic density and adjust the lights to the actual traffic load. These approaches are transparent to the users, as the energy consumption is taken into account automatically.

However, in order to really make a change towards green transportation we need to change the mind set of the users. One approach could be road pricing, where the tax is based on driven distance, number of people in the car and driving style. In order to keep the tax to be payed at a minimum people are encouraged to drive green, i.e. energy efficient, and take the environment into account. Here RFID again plays a role, since it can enhance the precision of the road pricing – e.g. the distances driven when the car has three passengers can be priced less than the distances driven when the car has a single passenger. RFID can be used to closely and securely log transportation data and thus be a key technology for introducing incentives towards achieving low-carbon transport.

8.5 Current State of RFID Usage

Even though RFID is already gaining foothold in areas like logistics and retailing the technology still have to be developed in order to utilize the full potential of RFID. As an example consider a large retailer like Wal-Mart, who has implemented RFID for inventorying when pallets arrive at each supermarket. This has proven beneficial for the supermarket itself, however, RFID holds the potential of being beneficial to the customers as well, in terms of easier checkout and more comprehensive information on the life cycle of each unique object in the supermarket. In order to make applications like that secure, reliable and cost-effective the research community still needs to solve some limitations in the RFID technology. These limitations are in this section introduced along with the current state of the art, in 2009, in each area.

8.5.1 Costs

One clear limitation of RFID systems is still the price per tag and to some extend the installation costs. This means that RFID systems is so far mainly used in applications where there is a reasonable ratio between tag and object price. In this chapter we focus on passive UHF tags, where prices so far have come down to around 7 to 15 US cents per tag, depending on the quantity and packaging of the tag, i.e. whether it is encased in plastic or embedded in a label [1].

In order to further reduce tag prices production costs must be brought down. The solution to cheaper tags is twofold. Research in chip and antenna design and production apparatus towards cheaper manufacturing and production of tags [30]. In order to drive, and probably also finance, this research large quantity demands for RFID from the industry is necessary, as mass production will contribute to a lower average price per tag [32]. However, demands for RFID and research towards cheaper tags are interdependent. Cheaper tags will increase the interest in, and demand for, RFID systems. Increased demand for RFID drives the research towards lower productions costs in order to increase profit, and due competition between tag manufactures, the free market will eventually cause the tag price to be reduced. Hence we need someone to initiate this development by taking the first step. Example of front-runners like this are Wal-Mart, Airbus and United States Department of Defence, which, due to their size, can afford the risk [2]. More initiatives like this will help fueling the development within RFID systems.

Moreover when integrating RFID systems we are facing the same considerations with respect to *clean* and *wean* slate design, as described in section 8.2.2. There will most probably be a difference in Return Of Investment, ROI, time when RFID systems and their possibilities are taken into account from the earliest phases when planning a new facility, compared to installation in an existing facility, where it may replace older equipment. The wean slate design may not be cost-effective at all, neither with respect to price nor carbon footprint.

8.5.2 Tag Antenna

In a passive UHF tag the antenna is essential as it harvests the energy to power up the tag. However, the antenna dimensions tend to be large compared to the desired tag dimensions, even when UHF antennas is considered. In order to fit into these dimensions the antenna may undergo different deformations, which influences the antenna characteristics [9]. Moreover, placing the tag on an object will have an effect on the antenna parameters as well, as the reflection and polarization coefficients may change significantly. This has the potential of causing the power received by the tag from the reader to change accordingly, which may cause false negative or false positive readings, respectively. These types of erroneous readings are described in detail in Section 8.5.3. This means that a manufacturer can develop a tag with an antenna that fulfills the requirements with respect to dimension, directivity and frequency. But when the tag is attached to an object of certain materials, e.g. liquid or metal, the antenna parameters are impacted to a degree that may render the tag completely useless. Knowledge of these effect is very important when designing the antenna for RFID tags. Unless different types of antennas is used for different objects materials, a cheap universal antenna needs to be developed, e.g. by shielding from the object, in order for RFID tags to become ubiquitous.

8.5.3 Erroneous Readings

If multiple tags are located in the proximity of a reader they will respond simultaneously to the reader and the reader experiences tag collision. Hence the protocol stack implemented at the reader, usually EPC Global Gen2 stack for passive UHF tags [12], should contain anti-collision protocols to resolve this collision [29]. In the ideal case, all the tags are identified, but only inside the readers interrogation zone, i.e. in a radius of approximately 3 m [9]. However, in real communication scenarios the reader can experience two types of erroneous readings, namely negative readings (or missed tags) and false positive readings. The details of these two types of erroneous readings and the current state of the art in order to cope with them are described in this section.

8.5.3.1 Missing Tags

A tag will not respond to a reader if the received power or Signal-to-Noise Ratio, SNR, is too low. This can occur even if the tag is located within the communication range of the reader, due to objects blocking the signal or the tag antenna has been deformed causing the gain in the direction of the reader to be too low to correctly receive the reader request. In this case the tag location is referred to as a *blind spot*. The problem of identifying tags in blind spots is called the *missing tag problem*. The case where a tag, supposedly in communication range, does not respond to a reader request is referred to as a False Negative reading, FN. As described in Section 8.5.2 the material of the object is also of importance as it can change the antenna parameters significantly. In [8] some typical identification rates are listed, and depending on the material the rate varies from 33.0 to 95.0%, for Chocolate Mousse and Rice Cake respectively.

Multiple approaches exist for decreasing the probability FNs due to tags located in blind spots. In [4] a method to increase the completeness of the interrogation is presented. This method is aimed at static tag populations and let each tag store a reference to another tag in the interrogation zone. The reader then compares the set of identified tags with the set of reference tags, i.e. the tags the identified tags refer to. If a reference tag is not in the set of identified tags the set is incomplete and the reader must interrogate again in order to identify the tag in the blind spot. In [8] two independent samples are used to estimate the cardinality of the tag set, with is then compared to the actual number of identified tags in order to determine if an additional reading is required. Another approach is utilized in [17], where in addition to estimating the tag cardinality, also estimates the number of required readings to guarantee, with some probability, that no tags are missing.

8.5.3.2 False Positive Readings

If a reader successfully receives a reply from a tag located outside its communication range it is referred to as a False Positive reading, FP. This type of erroneous reading is difficult to cope with as it is a valid response and therefore causes ambiguity about the presence/proximity of the tag [11].

In [20] it is stated that false positive readings occurs in approximately 5% of all readings, and in order to cope with these errors different measures have been presented. In [19, 18] deterministic data cleaning is utilized to sort out the FP readings. However, due the ambiguity of these false positives a deterministic model is not suited, hence in [20] a probabilistic model is proposed for cleaning the data. Another simpler approach, presented in [5]. The motivation of the approach is, since FNs occur due to the probabilistic nature of the wireless channel, then a simple way to decrease the number of false positives is to utilize these stochastic behavior. This approach defines an inventory request from a single reader as a window of N readings. A tag is said to be present in the interrogation zone only if it is identified in k out of N readings. For large N and k the probability of a tag being absent from the interrogation zone to be identified as being present becomes low.

This concludes the description of the problems and limitations in RFID systems, in the current state of the technology. When solutions to these obstacles has been developed, RFID systems can be ubiquitous and makes therefore a great candidate for an ICT that can be put to work in order to reduce GHG emissions.

8.6 Conclusion

In this chapter we have addressed the issue of using ICT systems to decrease emissions of greenhouse gases and thus reverse the global warming trend. The real green role of ICT, at least in a short term, can be seen if it is applied to the systems that bear the chief responsibility for increased carbon emissions, such as the power distribution grid, transportation, buildings, production processes, etc. In order to use the potential of ICT working towards a greener world, the deployment of wireless devices will explode, hence it is equally important to develop low power solutions and applications where energy efficiency is provided by ICT. We have identified RFID and, more general, the technologies based on passive wireless devices, as the ones holding large potential to facilitate low-carbon operation of the future systems. We have exemplified such usage of RFID through several applications. Finally we have described some problems and challenges for the RFID technology in its current state. These problems are identified as some of the main limiting factors with respect to developing innovative RFID applications and motivating market penetration for RFID systems. Hence we believe that the observations in this chapter will motivate research and innovation in the area of RFID and further studies on the green potential of RFID. Moreover, motivate the continuous development of low power solutions and innovative thinking regarding other green applications that rely on passive wireless devices.

References

[1] How much does an RFID tag cost today? From `http://www.rfidjournal.com/faq/20/85`.

[2] Industry leaders to explain benefits of RFID-provided visibility, 16 November 2009. From `http://www.rfidjournal.com/article/view/5383`.

[3] U.S. Department of Energy, *International Energy Outlook 2008*, 2009.

[4] Backes, M., Gross, T.R. and Karjoth, G., Tag identification system. U.S. Patent Office, August 2008.

[5] Bai, Y., Wang, F. and Liu, P., Efficiently filtering RFID data streams. In *Proceedings CleanDB Workshop*, 2006.

[6] Birkeland, L., Brunvoll, A., Hauge, G., Hoff, E. and Holm, M., How to combat global warming. Working paper prepared by the Energy and Climate Department of The Bellona Foundation, 2008. Available from: `www.bellona.org/reports_section`.

[7] Buettner, M., Greenstein, B., Sample, A., Smith, J.R. and Wetherall, D., Revisiting smart dust with RFID sensor networks. In *Proceedings Seventh ACM Workshop on Hot Topics in Networks (HotNets-VII)*, 6–7 October 2008.

[8] Weiss, L., Chaves, F., Buchmann, E. and Böhm, K., TagMark: Reliable estimations of RFID tags for business processes. *Proceeding of the 14th ACM SIGKDD International Conference on Knowledge Discovery and Data Mining, KDD'08*, 24–27 August 2008.

[9] Dobkin, D.M., *The RF in RFID: Passive UHF RFID in Practice*. Communication Engineering Series Elsevier, 2008.

[10] Eklund, J.M., Riisgaard Hansen, T., Sprinkle, J. and Sastry, S., Information technology for assisted living at home: Building a wireless infrastructure for assisted living. In *Proceedings Engineering and Medicine in Biology Conference*, 1–4 September 2005.

[11] Engels, D.W., On ghost reads in RFID systems. *White Paper from Auto-ID Labs, MIT, Camebridge, USA*, pp. 1–15, September 2005.

[12] EPCglobal, Specification for RFID Air Interface (1.1.0), 17 December 2005.

[13] Feist, W., Peper, S. and Gorg, M., Projectinformation, *CEPHEUS*, **38**, July 2001.

[14] Directorate-General for Research Sustainable Energy Systems. European SmartGrids Technology Platform – Vision and Strategy for Europe's Electricity Networks of the Future. 2006.

[15] Graham-Rowe, D., Wireless power harvesting for cell phones. *Technology Review*, June 2009. http://www.technologyreview.com/communications/22764/.

[16] Greenpeace, Why the G8 needs to finance Developing Country climate action, 2008.

[17] Jacobsen, R., Nielsen, K.F., Popovski, P. and Larsen, T., Reliable identification of RFID tags using multiple independent reader sessions. In *Proceedings of IEEE RFID 2009*, 27–28 April 2009.

[18] Jeffery, S.R., Alonso, G., Franklin, M.J., Hong, W. and Widom, J., Declarative support for sensor data cleaning. In *Pervasive 2006* Lecture Notes in Computer Science, Vol. 3968, Springer, pp. 83–100, 2006.

[19] Jeffery, S.R., Garofalakis, M. and Franklin, M.J., Adaptive cleaning for RFID data streams. *Proceedings of Conference on Very Large Data Bases (VLDB'06)*, pp. 163–174, 12–15 September 2006.

[20] Khoussainova, N., Balazinska, M. and Suciu, D., Probabilistic RFID data management. Technical Report TR2007-03-01, University of Washington, June 2007.

[21] Lee, C.K.M. and Chan, T.M., Development of RFID-based reverse logistics system. *Expert Systems with Applications*, **36**(5), 9299–9307, July 2009.

[22] Metz, B., Davidson, O., Bosch, P., Dave, D. and Meyer, L., Contribution of Working Group III. Fourth Assessment Report of the Intergovernmental Panel on Climate Change, 2007. Available from: http://www.ipcc.ch/ipccreports/ar4-wg3.htm.

[23] Nereng, G., Romsdal, A. and Brekke, A., Can innovations in the supply chain lead to reduction of GHG emissions from food products? A framework. In *Proceedings for Conference On Joint Actions on Climate Change*, 8–10 June 2009.

[24] Pamlin, D. and Pahlman, S., Outline for the first global IT strategy for CO2 reductions. In *Proceedings of the Future of the Internet Economy: ICT's and Environmental Challenges*, 2008.

[25] Paradiso, J.A. and Starner, T., Energy scavenging for mobile and wireless electronics. *IEEE Pervasive Computing* **4**(1), 18–27, 2005.

[26] Puccinelli, D. and Haenggi, M., Wireless sensor networks: Applications and challenges of ubiquitous sensing. *IEEE Circuits and Systems Magazine*, 2005.

[27] Massoud Amin, S.S. and Wollenberg, B.F., Toward a smart grid: Power delivery for the 21st century. *IEEE Power and Energy Magazine* **3**(5), 34–41, 2009.

[28] Sample, A.P., Yeager, D.J., Powledge, P.S., Mamishev, A.V. and Smith, J.R., Design of an RFID-based battery-free programmable sensing platform. *IEEE Transactions on Instrumentation and Measurement* **57**(11), 2608–2615, November 2008.

[29] Shih, D.H., Sun, P.-L., Yen, D.C. and Huang, S.-M., Taxonomy and survey of RFID anti-collision protocols. *Computer Communications* **29**, 17, 2006.

[30] Soldatos, J., AspireRFID can lower deployment costs. *RFID Journal*, 16 March 2009.

[31] Spinar, R., Muthukumaran, P., de Paz, R., Pesch, D., Song, W., Chaudhry, S.A., Sreenan, C.J., Jafer, E., O'Flynn, B., O'Donnell, J., Costa, A. and Keane, M., Efficient building management with IP-based wireless sensor network. In *Proceedings 6th European Conference on Wireless Sensor Networks*, 2009.

[32] TIBCO, Implementing RFID for rapid ROI and long-term success. White Paper, 2004.
[33] Urge-Vorsatz, D. and Novikova, A., Potentials and costs of carbon dioxide mitigation in the world's building. *Energy Policy*, **36**(2), 642–661, February 2008.

PART II

ENERGY/POWER OPTIMIZATION

9

Energy-Efficient Deployment through Optimizations in the Planning of ICT Networks

Rasmus Hjorth Nielsen, Albena Mihovska,
Ole Brun Madsen and Ramjee Prasad

Center for TeleInFrastruktur (CTIF), Aalborg University, 9220 Aalborg, Denmark

Abstract

The smoothless provisioning of broadband services to everyone is considered one of the key components of the information society. Traditionally, the delivery of broadband connections to the end user has been targeted through the deployment of optical fibers leading to the concepts of fiber to the curb and fiber to the home. At the same time, the impact of the wireless technology to provide a new sense of freedom and autonomy for the end user has gained its own momentum. This chapter proposes optimizations in the construction of information and communication technologies (ICT) networks, which are beneficial in a green ICT perspective as it saves components – active (for lower power consumption) as well as passive (lower production output). The considered scenario is an integrated fixed broadband access with wireless communications. An automated planning process can speed up the planning stage and lower the overall cost of the network deployment. This chapter is organized as follows. Section 9.1 gives an introduction into the topic. Section 9.2 describes the trends in the deployment of ICT networks. Section 9.3 proposes optimization strategies that can lead to a more energy-efficient deployment. Section 9.4 concludes the chapter.

Keywords: Green ICT, network planning, convergence, optimization.

R. Prasad et al. (Eds.), Towards Green ICT, 117–129.

9.1 Introduction

Green ICT is one of the main topics dealt with by governments and product makers. Novel ICT solutions are being proposed for support of the electricity networks (e.g., smart electricity distribution networks), as the basis for the development of tools leading to energy-efficient solutions in various spheres of the industrial and every-day life (e.g., monitoring and control systems able to optimize the local energy generation-consumption). This chapter focuses on the challenge of designing ICT networks and communications as more energy- and cost efficient. One way to achieve this is through optimizations of the design.

There are many aspects that can make ICT networks more efficient and environmentally friendly, while ensuring the same or larger capacity. Base stations powered by wind turbines are an approach to energy efficiency and sustainability for the wireless communication infrastructure [1]. Advanced technologies relying on distributed architectures are means to reduce the information exchange and release resources without the need for new deployments. Here, it should be mentioned that although relays can give the benefits of increased coverage and other benefits of a distributed architecture [2] a very dense deployment might raise environmental concerns. Network planning, including the reuse of existing infrastructure, therefore, is an essential aspect of developing energy-efficient and sustainable (e.g., 'green') communications.

The pattern of commuting in small cities and rural areas can be changed to benefit the overall reduction in CO_2 emissions by designing a digital infrastructure able to support a high quality of the communications and services for the home and the office. A high quality digital infrastructure enables the extensive use of home workspaces and videoconferencing thereby decreasing the commuting needs – local and global. Fiber to the home (FTTH) in combination with a reliable wireless network infrastructure is one enabler of high-quality ICT services [3].

9.2 Trends in the Deployment of ICT Networks

Optimizations in the design of ICT networks have mainly dealt with the aspect of obtaining cost minimized networks that are technically feasible [4, 5].

Site-sharing (i.e., the sharing of the physical premises – a rooftop, a mast in the countryside) is a current trend among operators. In the simplest case of

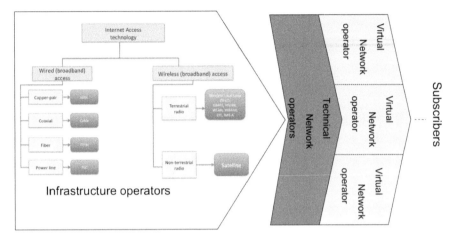

Figure 9.1 Future wireless ecosystem.

site-sharing, one operator lends to another operator a physical placement of its own at a given rent.

To prove the cost effectiveness of future mobile communications systems, suitable deployment cost modeling and analysis methodologies can be used [2]. Such procedures allow for analyzing the performance versus the costs trade-off for different technology options and for comparisons with already deployed technologies. The introduction of advanced radio technologies (e.g., flexible spectrum sharing) in a converging communication environment calls for investment decisions in a changing and complex wireless ecosystem. Figure 9.1 shows an example of a future wireless ecosystem.

The infrastructure operators own the converging communication infrastructure (e.g., fiber, WLAN, WiMAX, UMTS, broadcast systems, IMT-Advanced systems, etc.). The technical network operators rent the capacity in this infrastructure that is then combined in packages so that bit pipes fulfilling certain QoS requirements can be offered to virtual network operators who can sell the service to the subscribers.

For the design of ICT networks the objective of energy efficiency overlaps the cost optimizations in many ways. The cost of the network can be divided into the capital expenditures (CAPEX) and the operating expenses (OPEX) of operating it. CAPEX corresponds to the deployment cost of the network as a trade-off between the usages of elements in the network – active as well as passive. OPEX costs c_k per time period can be treated as discounted cash flows and represented by their present value (CAPEX) c at the beginning of the network deployment as shown in Equation (9.1):

$$c = \sum_{k=0}^{K-1} \frac{c_k}{(1+d)^k} \tag{9.1}$$

The planned lifetime K of the network (number of OPEX time periods) as well as a discount/interest rate d have to be determined beforehand.

The saving of elements during the build-up of the network has clear advantages as the energy used for production can be saved, while future obsolete equipment, which might have to be recycled or discarded can be limited. The OPEX corresponds to the operational cost of running and maintaining the network and savings on this can be obtained, e.g. from power savings in the equipment. An approach that can optimize already existing network planning strategies and can lead to an energy-efficient deployment at a lower cost was proposed in [6].

This chapter investigates the optimizations in the network planning process that can yield an environmentally friendly deployment of next generation communication technologies. In particular, a scenario is considered where fiber technology delivers the last mile access (FTTH).

9.3 Optimization Strategies for Energy-Efficient Deployment

The optimization problems addressing energy efficiency are many and the following will investigate some of the strategies for optimizing the network design where energy savings can be achieved.

9.3.1 Location of Facilities

The facility location problem minimizes the combined cost of deploying the facilities and connecting the customers to these. Facilities cover a broad range of sites from major houses of active equipment to street cabinets providing passive interconnection in wired networks and rooftops or mast for antennas in wireless networks. Deploying new facilities for green-field scenarios has environmental consequences in relation to the production of the facility and the operational issues in respect to the power consumption, cooling, etc., and the overall energy consumption can be accumulated in the same way as done for the investment where CAPEX and OPEX are considered. Limiting the number of facilities is beneficial from a power saving perspective but is a

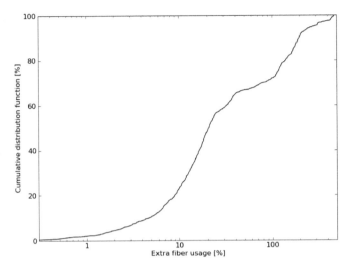

Figure 9.2 The distribution of extra fiber usage for different facility locations in a FTTH planning area.

trade-off with the power used for transmission especially relevant for base stations where the transmission power needed increases drastically with the distance between the base station and the user. Facility location problems also include the optimization in relation to brown-field scenarios when new facilities are being deployed. For such scenarios focus on energy efficiency can be increased, e.g. by including cell planning or more generally to consider the segmentation of the network with respect to users.

For wired as well as wireless ICT networks the optimal location of facilities does not only minimize the connection cost, but also optimize the performance and usage of the customer connections. For planning of facilities in future ICT networks these issues must be investigated.

For wired networks especially, the location of the facilities strongly influences the possible savings for the significant contribution of path planning. Figure 9.2 shows an example of the extra fiber usage compared to the minimum for different locations of a facility within a FTTH planning area. As seen the extra fiber usage varies greatly with the potential facility locations that are here the different households.

Similar plots can be done for the location of base stations and the increase in transmission power. The tradeoff between the energy usage of facilities and the energy usage related to the connection can be illustrated as in Fig-

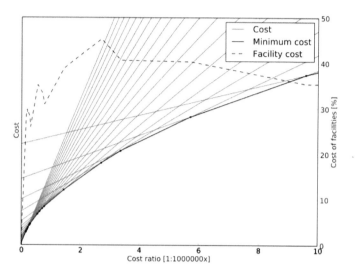

Figure 9.3 The choice between different numbers of facilities when considering the cost or energy ratio between a facility and the corresponding connection cost.

ure 9.3 where the costs of facilities and connections can be replaced by the corresponding energy usages. The figure shows the overall cost as the linear combinations of the number of facilities and length of connections needed for given cost ratios between the two parameters. To obtain energy minimized design the cost is replaced with energy.

9.3.2 Path Planning

Path planning is the task of interconnecting the elements in the network and includes the path between the range of facilities presented in the previous subsection and users as well as the interconnection of the facilities themselves, e.g. on different levels of the network.

Optimized planning of paths for deployment of fibers is another way to achieve power savings not only due to the saved materials in form of tubes and fibers or due to the saved construction time and energy, but also due to the decreased path loss and the possibility to lower the transmission power.

For existing network operators, the optimization related to path planning can consider the already deployed infrastructure for utilization or reuse where feasible.

For FTTH networks the main contribution to CO_2 is the deployment stage (80%) with the contribution from the passive fiber network estimated

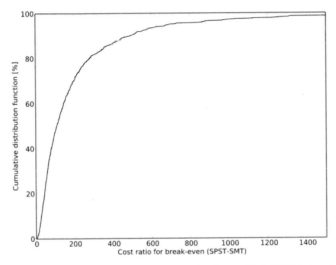

Figure 9.4 Choosing between path planning approaches for multiple FTTH planning areas for different costs of fiber and trenches.

to around 83% [7]. Optimized path planning is therefore much more than a minimization of the capital investment, but the two objectives are likely to overlap.

Optimization of path planning can be done on different objectives. A minimum fiber network design can be obtained by using shortest path spanning trees (SPST) to do the path planning. This is the traditional method for doing path planning in wired networks and the SPST can be found efficiently using e.g. Dijkstra's algorithm [12]. Another approach is to do path planning using the Steiner minimal tree (SMT) which leads to a minimum trench network design. Contrary to the SPST, there exists no efficient way to find a SMT as this problem is NP-complete [15], but due to extensive reduction frameworks (see e.g. [13]) problems can be reduced considerable and the SMT becomes solvable for many if not most real world problem scenarios. As an example the average reduction rate is more than 99% for interesting areas for FTTH planning [6]. Depending on the most energy consuming part of the deployment the design can be chosen from one of the two approaches. Figure 9.4 shows an example of the distribution of the break-even costs (the ratio between unit cost for trench and fiber) for the two approaches for different FTTH areas.

The break-even cost can either be the actual deployment cost or the energy consumption related to deploying trenches and fibers. For cost as well as en-

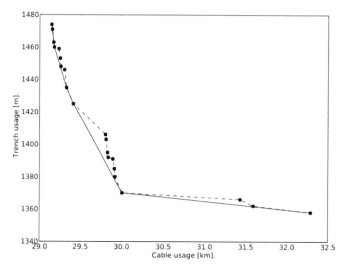

Figure 9.5 Choosing between different solutions with respect to the energy consumption related to cables and trenches respectively.

ergy the choice of path planning approach depends on the area of deployment. In sparsely populated areas, a minimum trench network found using the SMT is often desirable, while the decreased usage of fibers makes a minimum fiber network, found using the SPST, the best choice for densely populated areas.

A trade-off between the two objectives can also be found by considering the cable trench problem (CTP) [8]. Instead of minimization on lengths of trees the energy cost of cables and trenches can be used in order to obtain the most energy efficient network design. Similarly to the SMT, the CTP is a NP-complete problem and as only a limited reduction framework exists [6], the CTP is only solvable for smaller planning areas. By the use of a heuristic algorithm, solutions can be obtained which are better than those of the SPST and SMT for most cost or energy ratios between cables and trenches. Figure 9.5 shows different combinations of cable and trench usage for a smaller planning area and from these combinations and given energy consumption for cables and trenches respectively, the most energy efficient solution can be chosen.

9.3.3 Capacity Planning

The utilization of newly deployed or existing equipment is also an optimization problem to investigate for possible power savings. Related to this is

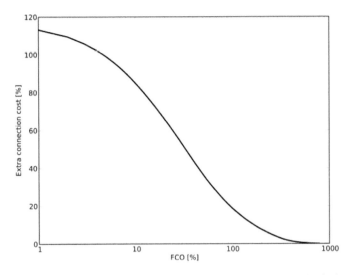

Figure 9.6 The increase in connection cost or energy as the extra capacity is decreased.

the capacity deployed in each facility and for wired networks this is a fixed design parameter defining the maximum number of customers that can be connected through a single facility. The choice of the appropriate capacity to install influences path planning as constraints on the capacity yields suboptimal solutions with respect to e.g. fiber in FTTH networks independent of the path planning approach. Furthermore the capacity is also related to the location of facilities as the installed capacity explicitly sets a lower bound on the number of facilities to deploy in order to cover the designated number of users.

Figure 9.6 shows the increase in connection cost or energy as the extra capacity (defined as the facility capacity overhead, FCO) is lowered.

Not utilized ports in active equipment are disadvantageous due to an increased power usage per customer and the excessive deployment of equipment that must be both produced and recycled.

Adding new base stations to the network should consider the cell planning in order to minimize the interference and thereby decrease the transmission power.

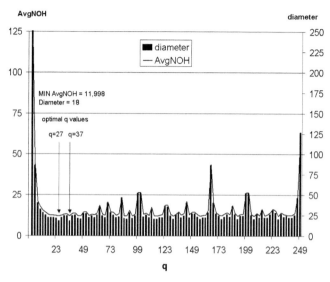

Figure 9.7 Performance of a 1000 node degree 3 network (N2R).

9.3.4 Structural Quality of Service (SQoS)

The interconnection of nodes in the backbone is also an issue for energy savings. Energy savings can be achieved by interconnecting nodes in structures with good SQoS parameters , hereby decreasing the average number of hops and the diameter of the network. The switching capacity in the nodes can be increased along with the power, while the network benefits from better performance and sustainability. Figure 9.7 shows the variation in performance depending on the number of nodes and their topology in a network.

9.3.5 Routing

The routing of packets in the network is also subject to optimizations increasing the energy efficiency. The more efficient a packet can be routed through the network, the less switching capacity is needed and energy consumption can be decreased. Overall it is desirable to lower the number of hops that a packet must taken in order to reach the destination. However, it should also be considered to include energy on links in order to choose the most energy efficient paths instead of simply the least hops one.

For this, unicast can be routed along the shortest path with respect to energy while broadcast can be achieved in the most cost and energy efficient way through a minimum spanning tree (MST). For multicast the most efficient paths can be achieved through the SMT as previously described.

More efficient routing schemes can be investigated by considering the topology of the underlying network structure with the use of SQoS [10]. By knowing the topology beforehand, routing can be done in a more efficient way while the inclusion of hierarchical layering to the topology can also lower the number of hops.

9.3.6 Combined Wired and Wireless Planning

By regarding the wired and wireless technologies as cooperating and not competing means of access, the design of the network can be done in the same manner [11]. The convergence between wired and wireless networks enables the full synergy effects in the deployment with shared locations, trenches, tubes or even fibers. However, for various reasons, a general and full quantification of the deployment of radio access systems can be impossible to perform. Certain deployment components are operator-specific as they depend, for example, on the contractual relationship between the hardware manufacturer and the operator. Others depend on the regulatory and legislative environment in the country of deployment.

A combined and optimized deployment planning can offer the benefits of savings in both the invested capital and the energy and CO_2 in the deployment stage. Furthermore, the user experience can be improved.

9.3.7 Multi Objective Optimization Problems

The previously proposed optimizations for energy efficient network design have all considered a single objective for which energy could be saved. The choice between cost and energy does not have to be either/or even in the cases where the two objectives do not overlap. To obtain such solutions multi objective optimization for the network design can achieve a deployment that benefits both the cost and energy criteria along with other criteria.

An example of a multi objective optimization problem with applications in both deployment and routing is the multi objective SPST [6]. This can be used for decision support for different deployment scenarios with cost and energy when these are related to the segments of the possible paths. To obtain all possible solutions might, however, not be feasible regardless of the efficient solution of the SPST as the multi objective problem is #P-complete [14] and the number of solutions can increase exponentially. The multi objective

SPST can also be used to support the choice of different paths (e.g., during routing in order to choose the most cost and energy efficient paths that still maintain the QoS guarantees).

Also for the other optimization problems in relation to network planning presented here, multi objective formulations can be done in order to include energy as a minimization criterion along with cost. However, many of the problems have a hardness of solution, which alter them difficult to solve in the single objective case and impossible in the multi objective case.

9.4 Conclusions

This chapter addressed the energy-efficient deployment of new ICT networks through optimization in the planning of the networks. New generation networks are likely to enable a decrease in energy due to the smoothless provisioning of new high quality services. However, the deployment of the networks must be energy efficient as well and in this chapter we have pointed out some of the most relevant topics and optimization problems to achieve this goal.

By including energy efficiency into the network design with respect to number and location of facilities, path planning, interconnection in the backbone, capacity planning and routing, significant savings in the energy and CO_2 are possible.

References

[1] Gozalves, J., White spaces, *IEEE Vehicular Technology Magazine* **4**(2), 2009.
[2] IST FP6 EU-funded Project WINNER, at www.ist-winner.org.
[3] ICT FP7 EU-funded Project FUTON, at www.http://www.ict-futon.eu/.
[4] Resende, M.G.C. and Pardalos, P.M. (Eds.), *Handbook of Optimization in Telecommunications*, Springer Science + Business Media, New York, 2006.
[5] Van Hoesel, S., Optimisations in telecommunication networks, Maastricht Research School of Economics of Technology and Organisation, 2004.
[6] Nielsen, R. H., Automated planning models for future ICT networks, PhD Thesis, 2009.
[7] Ecobilan for FTTH Council Europe, FTTH solutions for a sustainable development, http://www.ftthcouncil.eu/, 2008.
[8] Vasko, F.J., Barbieri, et al., The cable trench problem: combining the shortest path and minimum spanning tree problems, *Computers and Operations Research* **29**(5), 2002.
[9] Madsen, O.B., et al., SQoS as the base for next generation global infrastructure, in *Proceedings of IT&T Annual Conference*, 2003.
[10] Pedersen, J.M., et al., Applying 4-regular grid structures in large-scale access networks, *Computer Communications* **29**(9), 2006.

[11] Riaz, M.T., et al., A framework for planning a unified wired and wireless ICT infrastructure, *Wireless Personal Communications*, [online].

[12] Dijkstra, E.W., A note on two problems in connexion with graphs, *Numerische Mathematik* **1**, 1959.

[13] Hwang, F.K., Richards, D.S. and Winter, P., *The Steiner Tree Problem*, Annals of Discrete Mathematics, Vol. 53, North-Holland, 1992.

[14] Ehrgott, M., *Multicriteria Optimization*, 2nd ed., Springer, 2005.

[15] Karp, R.M., *Reducibility among Combinatorial Problems, Complexity of Computer Computations*, Plenum Press, 1972.

10

Low Power Hardware Platforms

Alberto Nannarelli

DTU Informatics, Technical University of Denmark, Kongens Lyngby, Denmark

Abstract

Over the past years low power design methodologies for hardware platforms have gained importance as a key factor to reduce the costs and improve the reliability of digital systems. The forecasts for energy consumption in the area of Information and Communication Technology (ICT) in the near future make the adoption of low power methods and techniques an unavoidable choice in the design of power efficient electronic systems.

Keywords: System-on-Chip, low power, energy consumption.

10.1 Introduction

At the beginning of the 1990s, the increasing number of transistors available on a chip made possible to integrate larger systems on a silicon die, bringing on the market a variety of portable products such as cellular phones, laptop computers, personal digital assistants (PDAs), GPS receivers, and medical devices. On the other hand, increased device densities on chip, consequence of the technology scaling, and faster clocks resulted in an increase of power dissipation that started to be a main concern in the design of integrated circuits.

The first effect of higher power dissipation in chips was the increase of costs in packages and cooling. A system that dissipates more than 2 W cannot be placed in a plastic package and the use of ceramic packaging, heat sinks, and coolant fans raises significantly the cost of the product. Moreover, the

R. Prasad et al. (Eds.), Towards Green ICT, 131–143.

chip functionality can be compromised in case of large current densities, since electromigration, caused by large currents flowing in narrow wires, might produce gaps or bridges in the power-rails of the chip with a consequent permanent damage of the system.

For portable systems, on the other hand, the critical resource is the battery lifetime, which can be lengthened by reducing the energy dissipation. This reduction also enables the use of smaller, and lighter, batteries.

Consequently, power consumption became in the 1990s an additional constraint in the design of integrated circuits, and methodologies and techniques for reducing the power dissipation started to emerge [4, 11, 13].

Today, the needs to reduce the energy consumption are even more compelling. For example, in [6] it is reported that the power necessary for server farms has doubled from about 60 TWh/year in 2000 to more than 120 TWh/year in 2005. Google's engineers are putting a lot of effort in making their warehouse-sized servers power efficient [5].

In [12], future estimations forecast that the amount of electrical energy necessary in 2020 to power ICT will be about 15% of the total. Considering that today 8% of the global electricity consumption goes in ICT, it is clear that building power efficient ICT equipment is a must.

In [20], a typical PC is taken as example and the energy necessary to manufacture it is compared to the energy necessary to operate the PC during the years it is in use. The comparison shows that the energy to manufacture the PC is about half the energy necessary to operate it for a few years. The conclusion is that extending the life time of electronic devices is beneficial for the energy budget.

In the next sections, an overview of the challenges in the design of systems implemented on silicon chips (hardware platforms) with respect to power dissipation is given. These chips are the heart of any electronic system: from high-performance computers to cellphones, from MP3 players to medical wearable devices.

10.2 Power Dissipation and Its Trends

10.2.1 CMOS Technology

The CMOS technology is the dominant one in the realization of integrated circuits and it is expected to stay dominant in the next decade. The complementarity (C of CMOS) of the transistors building a logic gate limits the current flow from power supply to ground (short-circuit currents) only for short periods when the gate switches. During the logic transitions the node capacitances of CMOS circuits are charged/discharged contributing to the dissipation of dynamic power. The well known formula characterizing the dynamic power dissipation in a CMOS logic gate, for example a NOT gate, is [19]

$$P_{dyn} = (C_L V_{DD}^2 + V_{DD} I_{sc} t_{sc}) f_p \qquad (10.1)$$

where V_{DD} is the power supply voltage, C_L is the capacitative load connected at the gate output, I_{sc} and t_{sc} are the short-circuit current and duration, respectively, and f_p is the toggling frequency of the gate. Expression (10.1) can be generalized for larger gates, containing several transistors and internal nodes, by introducing a term E^{int} – internal energy – which accounts for the short-circuit currents in all transistors and the power dissipated in the switching of the internal nodes. Expression (10.1) is then modified in

$$P_{dyn} = (C_L V_{DD}^2 + E^{int}) f_p \qquad (10.2)$$

However, CMOS transistors leak some current when not toggling. This component, called static power dissipation is

$$P_{stat} = V_{DD} I^{leak} \qquad (10.3)$$

where I^{leak} is the sum of all the leakage currents of the device: sub-threshold leakage, which is the drain-source current when the transistor is off; gate-oxide leakage, which is the current through the gate oxide (due to tunneling, etc.).

In synchronous systems composed by blocks of gates, the single gate toggling frequency f_p is normally related to the clock frequency f_{clk} by introducing the cell's switching activity defined as

$$a_i = \frac{n_i}{n_{clk}} = \frac{f_{pi}}{f_{clk}}$$

where n_i and n_{clk} are the number of transitions occurring in the i-th node (output of the i-th gate) and in the clock in a given time window.

Summarizing, by introducing the switching activity in (10.2) and combining it to (10.3), the total power dissipation in a CMOS cell part of a larger system is

$$P_i = (C_{Li} V_{DD}^2 + E_i^{int}) a_i f_{clk} + V_{DD} I_i^{leak} \qquad (10.4)$$

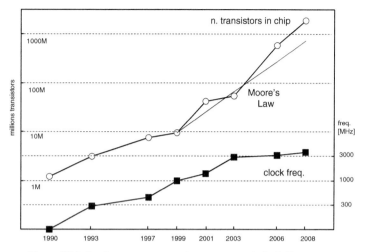

Figure 10.1 Intel processors transistor count and clock frequency.

10.2.2 Technology Scaling

Technology scaling is the main drive in microelectronics industry. As predicted by Gordon Moore, the number of devices packed in a chip doubles about every 18 months. As the transistors get smaller they get faster as well allowing for higher operating frequency. Dice size is also increasing across technology nodes, but at a lower rate than device scaling. As a result, the number of transistors in a chip increases every generation because of the device scaling and the increased die area.

Geometries are scaled by a factor S. The typical scaling factor is $S = \sqrt{2}$ from one generation (technology node) to the next. Together with the geometries, also the supply voltage is scaled by the same factor S. This is normally referred as the constant field technology scaling [19].

Under constant field scaling, for a system migrated to the new technology node the area is reduced by a factor S^2 (S in each dimension) and the delay reduced by S. Consequently, the dynamic power dissipation is reduced by S^2:

$$CV_{DD}^2 f \rightarrow \frac{1}{S} \cdot \frac{1}{S^2} \cdot S = \frac{1}{S^2}$$

However, the power density ($P/Area$) stays constant.

Figure 10.1 shows the trend for number of transistors (upper curve) and clock frequency (lower curve, right scale) for Intel's processor over the past two decades. The scale in the figure is logarithmic and shows that in the last

Table 10.1 Impact of leakage on technology scaling.

	P_{stat}	P_{tot}	$P_{stat}/P_{tot} \times 100$
mult (180 nm)	0.25	945	0.03
mult (120 nm)	2.25	450	0.50
mult (90 nm)	3.59	299	1.20
P_{90nm}/P_{180nm}	14.36	0.32	40.0

Power dissipation in μW at 100 MHz.

years the number of transistors on chip has grown more than the Moore's law predicts.

As for the clock frequency, Figure 10.1 shows a dramatic increase from 1997 (500 MHz) to 2003 (3 GHz), and then a plateau between 3–4 GHz in the past six years.

The main reason in this clock frequency trend is that the power dissipated would be so high to result in high costs for cooling down the chip and that the heat could compromise the functionality of the processor. For these reasons, the trend is now to increase the core parallelism (dual, quad, etc.) trying not to increase the chip power density.

10.2.3 Static Power Dissipation

With the technology scaling, and the increased transistor's leakage due to sub-threshold currents, also the static power dissipation starts to play an important role in today's power budgets.

To have an idea of the impact of the device's leakage on power dissipation, a multiply unit is synthesized in a 0.18 μm, a 0.12 μm and in a 90 nm library of standard cells. The same timing constraint, the delay of 25 inverters with fanout of 4 (FO4)[1] in their respective libraries is used in the synthesis. The results, shown in Table 10.1, indicate that the power dissipation due to leakage P_{stat} increases both in absolute value and as the percentage of the overall power dissipation P_{TOT}. By comparing the 0.18 μm and the 90 nm multipliers, P_{TOT} decreases of about 70% (mostly due to the scaling of V_{DD}), but the static part P_{stat} increases 14 times and its contribution to the total 40 times.

One way to reduce the static power due to leakage is to raise the MOSFET threshold voltage V_t. This can be obtained by changing the thickness of the oxide t_{ox}. However, a higher V_t has a negative impact on the device speed. To counter this problem, one solution is to have multiple cells implementing the

---

[1] FO4 is a standard measure of delay across different technologies.

Table 10.2 ITRS: three types of cells.

	ITRS2005 for 65 nm (2007)		
	L_g [nm]	t_{ox} [nm]	V_t [mV]
HP	25	1.1	165
LOP	32	1.2	285
LSTP	45	1.9	524

same logic function with devices with different V_t. This approach is similar to that used in the past to have several drive strengths. For this reason, the International Technology Roadmap for Semiconductors (ITRS) recommends the adoption of three types of cells for the same logic function (Table 10.2):

- High Performance (HP) for computers;
- Low Operating Power (LOP) for portable devices;
- Low Standby Power (LSTP) for portable and wearable devices.

The standard cell library used in the following to illustrate methodologies for power efficient design provides two classes of cells: cells with devices with a reduced threshold voltage designed to achieve high speed, identified as HS, and cells with devices with a higher V_t to provide low leakage identified as LL [16]. The LL cells are designed to have a I^{leak} several times smaller than the corresponding HS cells, but also the dynamic part is reduced with respect to the HS corresponding cell.[2]

10.2.4 Thermal Management

High power dissipation might result in excessive heating. Because silicon is not a good heat conductor *hot-spots* might form on the die in areas with high power density.

One of the consequence of increased die temperature is an increased leakage power that contributes to the rise of temperature in the hot-spot. This is clearly a vicious circle

$$T \uparrow \longrightarrow P_{leak} \uparrow$$
$$\searrow \quad P_{TOT} \uparrow \quad \nearrow$$

that should be avoided by distributing the hot-spots as further away as possible.

---

[2] A detailed explanation is given in [3].

Therefore, power and thermal management is becoming a very important area. Power and thermal profiling of the chip is the key to obtain an optimal placement of the most power hungry blocks, so that the chip temperature will not raise so high to compromise the correct system operations.

10.3 Low Power Design Methodologies

Expression (10.4) is a good starting point to model the contributions to the power dissipation. Expression (10.4) can be generalized for a system containing N cells

$$P_{TOT} = f_{clk} \underbrace{\sum_{i=1}^{N} \left(V_{DD}^2 C_{Li} + E_i^{int} \right) a_i}_{\text{dynamic}} + V_{DD} \underbrace{\sum_{i=1}^{N} I_i^{leak}}_{\text{static}} \qquad (10.5)$$

The quantities V_{DD}, E_i^{int} and I_i^{leak} depend on the technology and the gate (the latter two) used, C_{Li} depends on the design and layout (type of cell connected and wire capacitance), and f_{clk} and a_i depend on the design constraints and on the input pattern used for the specific application.

To obtain lower power dissipation and a more energy efficient design, one or more quantities in (10.5) must be minimized. Here some trade-offs are reported:

- By reducing the number of cells N, the area is reduced as well. However, usually additional cells are included to reduce the delay (e.g. by increasing parallelism), and eliminating those cells might slow down the system or, even worse, produce faults.
- Because of the quadratic dependency of dynamic power on V_{DD}, a reduction of the supply voltage produces significant energy savings. Unfortunately, the delay of a CMOS gate depends on supply voltage as well, and a reduction of V_{DD} impacts the system performance. However, for groups of cells not in the critical path, V_{DD} can be reduced to a value such that the delay in the corresponding path does not exceed the critical one.
- Reducing the nodes' load C_{Li} is very good because also the delay decreases. However, as in the case of N, C_{Li} is often due to the presence of gates which implement functions necessary for the system and cannot be eliminated. The term C_{Li} includes the wiring capacitance as well, and

this contribution can be reduced to the minimum necessary by accurate placement of cells and routing of the wires.

- The internal energy E_i^{int} is mainly due to short-circuit currents and charging/discharging of internal capacitance (including the internal clocking network for latches and flip-flops). While for the short-circuit contribution a design with rapid transition times tends to minimize the energy dissipation, for the other contributions, nothing can be done at design time when standard cells are used, as in most cases. Similarly, the contribution I_i^{leak} depends on the cell layout and cannot be reduced at design time.
- The reduction of the number of transitions, or the switching activity a_i, does not impact delay or area, in most cases. Therefore, reducing the switching activity is considered a very effective way of implementing power efficient systems.

Over the years a plethora of low power methods and techniques have been introduced. In the following, the most effective and promising are presented.

10.3.1 Reducing the Switching Activity

Switching activity, or transitions, can be reduced in different ways. Two of the most efficient ways are:

Changing the data encodings
A considerable amount of energy is dissipated in processor buses where capacitances are several orders of magnitudes larger than those of the internal nodes of a circuit [8]. To reduce the number of transitions in buses (and the power dissipation), different encodings of the binary vectors can be applied. Some examples of these encodings are given in [1, 15].

In signal processing if samples are represented in two's complement and are close to zero, several bits (the most significant ones) might flip (0 to 1, and vice-versa) if negative and positive numbers are alternating. The use of different numerical representations, such as the Logarithmic or the Residue Number System, proved to be beneficial for reducing the switching activity and, consequently, the power dissipation [3, 17].

Clock gating
By disabling the clock in flip-flops (FFs) which do not change their state in a given time window, energy can be saved not only in the FFs, but also in the logic connected to their outputs. Different methodologies can be

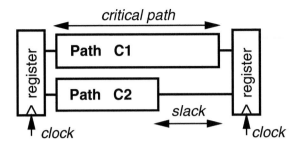

Figure 10.2 Slack in timing.

used to implement clock gating [2, 7]. Because the clock is distributed by a clock-tree, power management of blocks can be easily implemented by disabling the part of the tree not used at the moment.

10.3.2 Techniques to Reduce Both Dynamic and Static Power Dissipation

Figure 10.2 illustrates an example in which two paths register-register (C1 and C2) have different delays. One can take advantage by the delay difference (slack) to re-design the logic C2 for low power. This can be achieved, for example, by lowering V_{DD} to V_2 such that $t_{C1}(V_{DD}) = t_{C2}(V_2)$. In this way both the dynamic (quadratically) and the static (linearly) power dissipation are reduced in path C2. However, this method presents a few problems: (1) several voltage regulators have to be used to have an accurate delay matching and a consequent power saving; (2) voltage level shifters are required when going from the lower voltage to V_{DD} [18]; (3) having several power supply voltages complicates terribly the power grid. For these reasons, the method of reducing V_{DD} in non-critical paths is not practical in many cases.

On the other hand, the use of library with HS and LL cells can reduce the power dissipation without the drawbacks listed above. The method is similar: HS cells are replaced by LL cells in path C2 until $t_{C2}(HS + LL) \leq t_{C1}(HS)$.

This design technique is supported by logic synthesizers, so that it has little impact on the design flow. However, sometimes slacks are quite small and the benefits marginal. In the latter case, it might be useful to apply some circuit transformation to make slacks larger.

By retiming it is possible to create larger slacks and limit the critical path to a smaller portion of the circuit, as shown for example in Figure 10.3. Re-

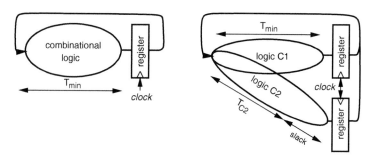

Figure 10.3 Original circuit (left). Retimed circuit (right).

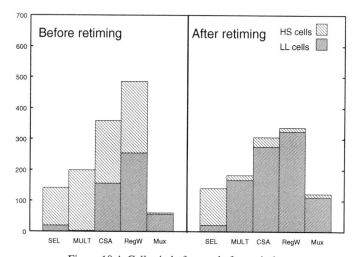

Figure 10.4 Cell mix before and after retiming.

timing is the circuit transformation that consists in re-positioning the registers in a sequential circuit without modifying its external behavior [9].

By applying retiming to a numerical processor, described in [10], power savings of about 30% are obtained, and the cell mix (HS, LL) is shown in Figure 10.4.

10.3.3 Leakage Control

Table 10.1 shows that although static power is increasing with technology scaling, it is still a small portion of the total compared to the dynamic part. However, in case the unit is not utilized often, the static contribution can exceed the dynamic one. Power gating is introduced to reduce drastically

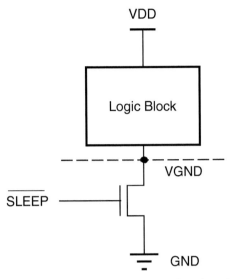

Figure 10.5 Power gating by sleep transistor insertion.

both the dynamic and static power when a unit (or a cluster of cells) is not used.

Power gating can be implemented by inserting a high-threshold NMOS transistor between the source terminal of the cell/block and the ground rail, as shown in Figure 10.5. This operation is called Sleep Transistor Insertion (STI). The insertion point of the sleep transistor is called Virtual Ground line (VGND).

The STI can be either individual (cell by cell), or clustered [14]. Clustered STI has a lower area overhead, and results in a lower congestion for the routing of the sleep signals (this operation is done post-placement), and lower buffering overhead to drive sleep signals.

When applying power gating a number of design issues must be taken into account:

- STI is a post-layout step, so it is important to produce a little perturbation of the original layout to achieve the final design closure.
- In sizing the Sleep Transistor (ST) the following trade-offs must be considered
    - The ST is up-sized to minimize the virtual ground voltage (V_{VGND}) and to produce the minimal timing overhead.

- For large sleep transistors the overall ST area is increased and the buffers result in long power-mode transition times (turn on/off times). Moreover, a non-negligible amount of energy is spent during the power-mode transition.
- Depending on the timing of the design and the ratio stand-by/active time. the size of the STs is chosen to achieve the optimal power savings.

10.4 Conclusions

In the past years the issue of power dissipation in silicon chips has grown of importance and it is currently probably the most pressing design constraint in digital electronics.

Over the years, a number of design methodologies and techniques have been developed to have power efficient systems. Today, design for low power is considered a standard practice in any hardware design flow.

The adoption of these methodologies enabled the portable electronics revolution (smartphones, MP3 players, etc.) and allowed to put a cap on the power consumption of high-performance computers. However, as the number of users of ICT is growing at an exponential rate, the effort to design low energy electronics must be intensified to keep the global energy consumption from growing too much.

In the near future, low power methodologies and power and thermal management are going to be integrated in the design flows of systems on silicon since the design early stages, and they will be in the *virtual toolbox* of every hardware engineer.

References

[1] Benini, L., De Micheli, G., Macii, E., Sciuto, D. and Silvano, C., Asymptotic zero-transition activity encoding for address busses in low-power microprocessor-based systems. In *Proc. of IEEE Great Lakes Symposium on VLSI*, pp. 77–82, March 1997.

[2] Benini, L., Siegel, P. and De Micheli, G., Automatic synthesis of gated clocks for power reduction in sequential circuits. *IEEE Design and Test of Computers*, pp. 32–40, December 1994.

[3] Cardarilli, G., Del Re, A., Nannarelli, A. and Re, M., Low power and low leakage implementation of RNS FIR filters. In *Proceedings of 39th Asilomar Conference on Signals, Systems, and Computers*, pp. 1620–1624, October 2005.

[4] Chandrakasan, A.P. and Brodersen, R.W., *Low Power Digital CMOS Design*, Kluwer Academic Publishers, 1995.

[5] Fan, X., Weber, W.-D. and Barroso, L.A., Power provisioning for a warehouse-sized computer. In *Proceedings of ACM International Symposium on Computer Architecture*, June 2007.

[6] Fettweis, G.P., ICT energy consumption – Trends and challenges. Presentation at ICT4EE: High Level Event on ICT for Energy Efficiency, March 2009.

[7] Lang, T., Musoll, E. and Cortadella, J., Individual flip-flops with gated clocks for low-power datapaths. In *IEEE Transactions on Circuits and Systems*, June 1997.

[8] Macii, E., Pedram, M. and Somenzi, F., High-level power modeling, estimation and optimization. *Proceedings of 34th Design Automation Conference*, pp. 504–511, June 1997.

[9] Monteiro, J., Devadas, S. and Ghosh, A., Retiming sequential circuits for low power. *Proceedings of 1993 International Conference on Computer-Aided Design (ICCAD)*, pp. 398–402, November 1993.

[10] Nannarelli, A. and Lang, T., Low-power divider. *IEEE Transactions on Computers* **54**, 2–14, January 1999.

[11] Nebel, W. and Mermet, J. (Eds.), *Low Power Design in Deep Submicron Electronics*. Kluwer Academic Publishers, 1997.

[12] Pickavet, M., Is ICT green? Presentation at Panel on Green Communication, IEEE International Conference on Communications, June 2009.

[13] Rabaey, J.M., Pedram, M., et al. *Low Power Design Methodologies*. Kluwer Academic Publishers, 1996.

[14] Sathanur, A., Calimera, A., Pullini, A., Benini, L., Macii, A., Macii, E. and Poncino, M., On quantifying the figures of merit of power-gating for leakage power minimization in nanometer CMOS circuits. In *Proceedings of IEEE International Symposium on Circuits and Systems (ISCAS 2008)*, pp. 2761–2764, May 2008.

[15] Stan, M.R. and Burleson, W.P., Bus-invert coding for low-power. *IEEE Transactions on VLSI Systems* **3**(1), 49–58, 1995.

[16] STMicroelectronics. 90nm CMOS090 Design Platform. http://www.st.com/stonline/prodpres/dedicate/soc/asic/90plat.htm.

[17] Stouraitis, T. and Paliouras, V., Considering the alternatives in low-power design. *IEEE Circuits and Devices Magazine* **17**, 22–29, July 2001.

[18] Usami, K. and Horowitz, M., Clustered voltage scaling technique for low-power design. In *Proceedings of International Symposium on Low Power Design*, pp. 3–8, April 1995.

[19] Weste, N.H.E. and Eshraghian, K., *Principles of CMOS VLSI Design*, 2nd edition. Addison-Wesley Publishing Company, 1993.

[20] Williams, E., Energy intensity of computer manufacturing: Hybrid assessment combining process and economic input-output methods. *Environmental Science & Technology* **38**(2), 6166–6174, November 2004.

11

Alternative Power Supplies for Wireless and Sensor Networks and Their Applications

Nobuo Nakajima

Department of Human Communication, Faculty of Electro-Communications,
University of Electro-Communications, Tokyo 182-8585, Japan

Abstract

Wireless and sensor networks are indispensable for "Green Communications". Lower cost and lower power consumption capabilities are important for these networks to achieve "Green Communications". A problem may be raised for these networks regarding to the electrical power supply, when the network spreads very wide area. Remote power feed technology is one of the solutions for this problem. In this chapter, three kinds of novel electrical power supply technologies are presented. They are (1) optical power feed through fiber, (2) optical power feed through air, and (3) fluorescent lamp power feed. Experiments show that they can supply several tens mW electrical power for the equipment. Examples of the applications are also presented.

Keywords: Power supply, optical fiber, sensor network, fluorescent lamp.

11.1 Introduction

"Sustainable Society", where the carbon emission and energy consumption are controlled as low as possible, is the most urgent issue on the earth. "Green Communications" are expected to play very important role for realizing this sustainable society.

R. Prasad et al. (Eds.), Towards Green ICT, 145–156.

Means to reduce carbon emissions are to minimize the consuming power and to improve the productivity in the industries, agricultures and human activities. Sensor networks will contribute for these purposes. For example, the sensor network monitors the temperature and reduces unnecessary power consumption of the air conditioners in the "smart building". By sensing the temperature variation in the farm and utilizing the result for farming, the productivity of the harvest will be improved. As a result, the carbon emissions can be substantially reduced because the additional energy consuming work is not needed to fulfil the target of the harvest.

The sensor network is composed of the sensors and a wireless network. One of the problems of the sensor network is how to get electrical power supplies. It is not easy to get the electrical power in the open area such as a farm and sometimes even in the indoor area.

For solving this problem, battery is one of the most useful alternatives. However, since the electrical energy of the battery is limited, charging or replacement is necessary. Furthermore, since some of the materials of the battery are harmful for plants and animals, the batteries must be collected after expired. Charging, replacement and collection consume energy. Remote power feed technologies proposed in this paper will be useful to overcome these problems.

In addition to the sensor networks, future mobile communication systems and ubiquitous communication systems are also useful to reduce the carbon emissions, since they are effective to improve the efficiency of the industries and human activities. However, these systems also have the power supply problem because base stations and ubiquitous nodes must be equipped wherever they are necessary.

Remote power feed technologies are useful for these issues, too. Figure 11.1 shows examples of alternative electrical power resources [1]. In this paper, three kinds of other electrical power supply technologies are presented: (1) optical power feed through fiber, (2) optical power feed through air, and (3) fluorescent lamp power feed. Applications of these power supplies are also presented.

11.2 Optical Power Feed through Fiber

Since the loss of optical fiber is extremely low, power can be efficiently sent through the fiber in the form of optical energy. The optical fiber power transmission has another advantage. The fiber cable is applicable even in the very strong magnetic field as well as in the water, salty or acid environments.

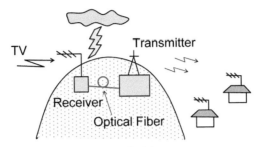

Figure 11.1 Television repeater.

Inexpensive 800–1060 nm wavelength laser diodes (LD) and efficient photo diodes (PD) with high power handling capability are used. More than 1 W output power laser diode is available for this purpose.

The optical fiber power transmission was already utilized for various purposes. In the case of broadcasting television repeater, the fiber sends power from a transmitter side to a receiver side in order to drive the receiver amplifier and electro-optical transducer (Figure 11.1). The purpose to feed the power by the optical power is to electrically isolate the receiver from the transmitter for preventing thunder attacks. The system written by Haeiwa et al. [2] uses 300 mW laser and 30 mW power is obtained after 6 km optical fiber.

The fiber power feeder is applied for supplying electrical power to the video camera [3]. Two graded index fibers are used and 1 W laser is connected to each fibers. As a result, 400 mW power is obtained for the video camera.

In this paper, the optical fiber is used to feed the power to the wireless repeater [4]. Figure 11.2 shows the basic structure of a proposed wireless repeater system. A base station and the repeater station are connected by the fiber. There are three optical channels with different wavelengths, they are forward link, reverse link and power transmission. These three channels are combined by multiplexers and transmitted through single optical fiber. The optical energy is fed to the amplifiers and a laser diode of the repeater station.

Figure 11.3 shows a laser diode (AF4A131675L made by Anritzu) and high power photo diode (PPC-9LW-SC made by NEL) of an experimental system. Figure 11.4 shows a relationship among output power, load resistance and available power. Adjusting the load resistance to achieve maximum available power, 47 mW was obtained from the photo diode when the output power of the laser diode is 300 mW. The efficiency was 16%.

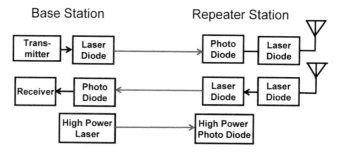

Figure 11.2 Basic structure of the fiber power feed.

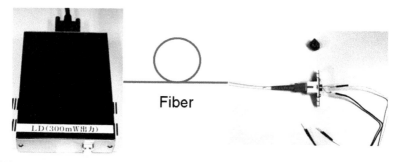

Wavelength: 1480 nm   Efficiency: Max 25%
Output Power: Max 300 mW   Output Power: Max 100 mW

Figure 11.3 High power laser and efficient photo diode.

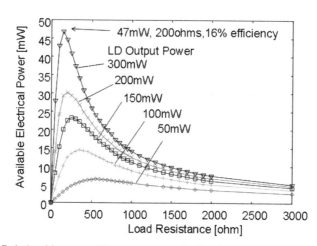

Figure 11.4 Relationship among LD output power, load resistance and available power at PD.

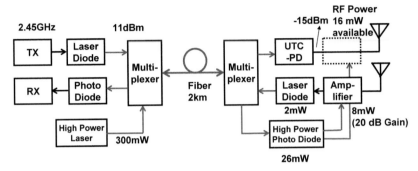

Figure 11.5 Wireless LAN repeater system.

Figure 11.5 shows a wireless LAN application operating at 2.4 GHz band. The level diagram is also indicated. A 300 mW laser diode is used for the power transmission. The power loss by the optical components and E/O transducers were about 10.6 dB. Measured available power at a high power photo diode was 26 mW. Since receiving amplifier and the laser diode laser consumes 8 and 2 mW, respectively, 16 mW still remains for a transmitter power amplifier.

11.3 Optical Power Feed through Air

If there is neither commercial power supply nor optical fiber at the wireless node, a wireless optical transmission is another alternative as the communication link and power source. A high power laser is applicable for the power transmission, but it has a problem for eye safety which is indispensable for commercial usage. Therefore, a conventional high intensity halogen lamp and LED are investigated [5].

Figure 11.6 shows an image of the wireless optical transmission system. A large lens reduces transmission loss. But it is not acceptable for actual use. Therefore, efficient power transmission is important. Available electric power is expected to be at least several tens mW.

Figure 11.7 shows the optical components for the power transmission. A high power LED and a tungsten halogen bulb are candidates of the optical power source. In this system, the tungsten halogen bulb is employed because of the higher output power than the LED.

In order to converge the high power light to a sharp beam, a lens is not suitable because it collects small portion of optical energy radiated from an

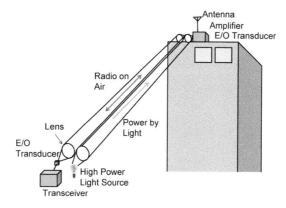

Figure 11.6 Image of optical wireless system.

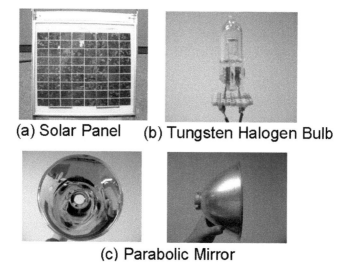

Figure 11.7 Optical components for power transmission.

optical source because the radiation is isotropic. A parabolic reflector can efficiently collect the radiated power than the lens as shown in Figure 11.8.

The optical beam diffuses even lens or the reflector is used. This diffusion is caused by the filament structure of the bulb. If the filament were a point source (infinitely small), the diffusion would not happen. The diameter of the optical beam at distance L from the transmitter is approximately $h = (L \times d)/f$ as shown in Figure 11.9, where d and f are filament and focal lengths, respectively.

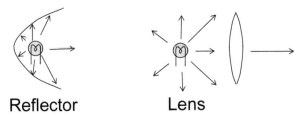

Figure 11.8 Optical radiation converging devices.

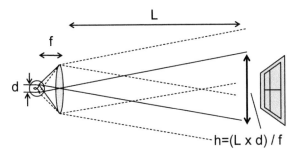

Figure 11.9 Diffusion caused by the length *d* of the filament.

Available electrical power at the receiver can be calculated under the condition of light source power, reflector/solar panel dimension, distance and solar panel efficiency. Figure 11.10 shows the theoretical result. An experimental result is also indicated in Figure 11.10. The experimental beam size is 174 cm × 65 cm on the solar panel. Figure 11.11 shows a photo of the parabolic mirror and the tungsten halogen bulb.

Figure 11.12 shows an example of the link budget for the wireless LAN repeater system using the proposed wireless optical power feed. By using 26 cm diameter reflector and solar panel, 10 mW RF output power is available at 40 m distance separation between the access point and the repeater station.

11.4 Fluorescent Lamp Feed

In the indoor environment, it is not so difficult to get commercial power supply comparing outdoor environments. However, some of the multi-hop wireless nodes might need to be equipped at the locations where the commercial power supply is not available. Since fluorescent lamps are equipped almost everywhere in the building, the fluorescent lamp feed can solve this problem as well as contributes the node installation cost reduction.

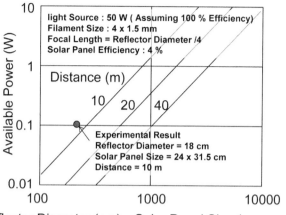

Figure 11.10 Available power vs. transmission conditions.

Figure 11.11 Tungsten halogen bulb with parabolic mirror.

Figures 11.13 and 11.14 show the structure and the photograph of the fluorescent lamp feed [6]. The principle is described as follows. Alternate current, of which frequency is several tens kHz, flows within the fluorescent lamp. If the tube is inserted in a toroidal ferrite core, an alternate magnetic flux flows in the core and it generates the electrical current in the coil that is rolled around the core. Finally, a DC power is obtained by rectifying the alternate current.

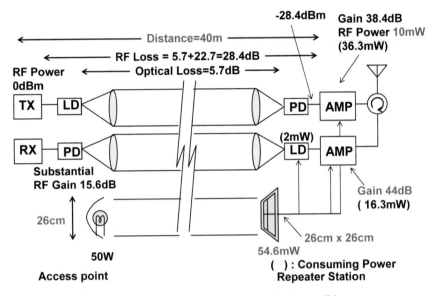

Figure 11.12 Available power vs. transmission conditions.

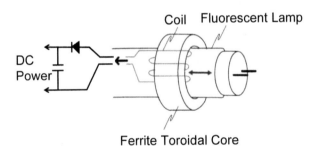

Figure 11.13 Fluorescent lamp power supply.

In order to obtain the maximum electrical power, a capacitor is inserted in the circuit to cancel the reactance of the coil (Figure 11.15).

In the experiment, since reactance of the coil was 14 mH and inverter frequency was 60 kHz, the calculated capacity becomes 500 pF. A value of the load resistance is also important. Figure 11.16 shows the relationship between the load resistance and the available power. When the load resistance is 300 ohms, 80 mW electrical power was obtained.

This power source was connected to a 315 MHz transmitter for range free positioning (Figure 11.17). The location data (latitude and longitude) is

Figure 11.14 Toroidal core with coil.

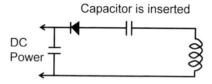

Figure 11.15 Fluorescent lamp feed circuit.

transmitted intermittently. The output power is 10 dBm and transmission bit rate is 2400 baud. Consuming power was 47.5 mW which is low enough for the fluorescent lamp power supply to feed.

The fluorescent lamp feed was also developed by NEC and the electric power of 50 to 100 mW is available.

11.5 Conclusions

Three kinds of the alternative electrical power supply systems are presented. These system can feed several tens mW power for sensors and wireless nodes. The fiber feed type is applicable for wireless LAN access points or base

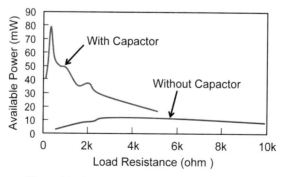

Figure 11.16 Available power vs. load resistance.

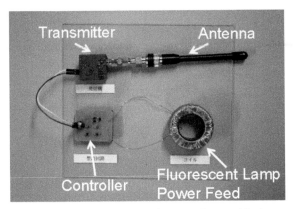

Figure 11.17 RF transmitter for range free positioning.

stations at the locations where the commercial power supply is not available. This technology will be especially suitable for femto cell base stations because the femto cell consumes relatively lower power and will be equipped everywhere regardless whether commercial power supply is available or not.

The optical wireless feed will be useful for outdoor sensor and wireless networks. A solar battery will be applicable. However, it does not operate during rainy or cloudy days. The proposed system can operate throughout the year. It can feed solar battery when rainy or cloudy days continue for a long time.

The fluorescent lamp feed system is useful for indoor wide area networks. If a number of sensors or wireless nodes are necessary, the installation cost becomes huge. The sensor or wireless node with this power feed system can be equipped only by inserting the fluorescent lamp into the toroidal core. Installation cost is negligibly small.

These power feed systems will be useful for future cellular communication systems, ubiquitous systems and sensor networks in the outdoor and indoor area for saving the energy, materials and cost of the installation.

References

[1] Texas Instruments, Energy harvesting, White Paper, http://focus.ti.com/lit/wp/slyy018/slyy018.pdf, November 2008.
[2] Haeiwa, Yamashita and Toba, Recent trends of light microwave fused technology in broadcasting, *IEICE Magazine*, Vol. 88, No. 9, 2005, pp. 735–739.
[3] http://www.icron.com/.

[4] N. Nakajima, ROF technologies applied for cellular and wireless systems, MWP2005, Seoul, Korea, October 2005.

[5] N. Nakajima and N. Yokota, Cellular/wireless LAN repeater system by wireless optical link with optical power supply, *WSEAS Transactions on Communications*, Vol. 7, No. 8, August 2008, pp. 882–891.

[6] T. Hazugawa, N. Nakajima and K. Hattori, Indoor wireless positioning system featuring lower installation cost, in *ASIA GIS 2008*, Session 11 LBS and Telematiques, September 2008, p. 66.

12

Energy Saving in Wireless Access Networks

L. Jorguseski[1], R. Litjens[2], J. Oostveen[1] and H. Zhang[1]

[1]*Access Network Technologies,* [2]*Performance Planning and Quality,*
TNO Information and Communication Technology, P.O. Box 5050,
2600 GB Delft, The Netherlands

Abstract

The ICT sector is responsible for 2 to 3% of the global energy usage, which is roughly equal to the energy consumption of all air traffic. Because of the fast growth of ICT usage worldwide, the share of the ICT sector in the overall energy budget is growing. Therefore, the sector is undertaking a wide set of activities to improve its energy efficiency and carbon footprint. Also the wireless communications industry is actively working in this area. This chapter presents the various activities at research institutes, vendors, operators, standardization bodies, regulatory/governmental bodies, etc. in reducing the energy consumption and carbon footprint of the wireless industry. It also presents an example of reducing energy consumption in wireless access networks via dynamically switching on/off base stations (and cells) according to the current traffic demands. A quantitative analysis is presented to estimate the benefits of this approach. The analysis show that dynamic base station (cell) switching on/off has significant potential for energy saving.

Keywords: Energy saving, wireless access, dynamic switching on/off.

12.1 Introduction

Reducing the extent of climate change is one of several huge challenges the world is currently facing. In December 2009, world leaders convened

R. Prasad et al. (Eds.), Towards Green ICT, 157–184.

in Copenhagen with the objective to agree on a set of effective measures to reduce carbon emissions in the coming decades. On a smaller scale, many governments, industry groups and companies are working towards a more energy efficient way of production and consumption. The ICT sector is responsible for 2 to 3% of the global energy usage, which is roughly equal to the energy consumption of all air traffic. Because of the fast growth of ICT usage worldwide, the share of the ICT sector in the overall energy budget is growing. Therefore, the sector is undertaking a wide set of activities to improve the energy efficiency and carbon footprint of its services. Moreover, the ICT sector plays a key role in developing ICT-based solutions for energy reduction in many other sectors.

Also the wireless communications industry is actively working to reduce its carbon footprint. Activities in the wireless industry include the use of renewable energy sources, the development of devices and infrastructure with lower energy consumption, and energy efficient operation of wireless access networks. In addition, organizations are defining targets and programs for energy reduction and energy efficiency. The most relevant global initiative is the publication of the 'Mobile Green Manifesto' [1] by the GSM Association (GSMA) [2], the mobile network industry organization. This manifesto sets out how the mobile industry plans to lower its greenhouse gas emissions per connection, and demonstrates the key role that mobile communications can play in lowering emissions in other sectors and industries. It also makes specific policy recommendations for governments and the United Nations Climate Change Conference in Copenhagen, including the 15th Conference of the Parties (COP15), in order to realize the full potential of mobile communications' ability to enable reductions in global greenhouse emissions.

This chapter presents an example of the possibility to reduce the energy consumption in the ICT sector. More precisely, we propose to use intelligent ICT components to improve the energy efficiency of wireless access networks.

Within 3rd Generation Partnership Project (3GPP) [3], the standardization body in charge of, e.g., UMTS/HSPA and LTE standardization, there is strong momentum to reduce the operational cost of mobile networks. This is achieved, among other approaches, by the introduction of 'self organization' solutions, i.e. advanced feedback and optimization/configuration mechanisms which reduce the need for manual network management tasks (configuration management, performance management, fault management). One such mechanism aims to introduce self-organized energy savings management.

The approach is to (automatically) switch off network resources which are not necessary to accommodate all network traffic at the required service quality level, and switch resources back on again once required to serve an increased traffic demand. As part of this chapter, we present a quantitative analysis of the potential energy gains that can be achieved from applying automated mechanisms for switching on/off capacity depending on the traffic demand.

This chapter is structured as follows. In the following section we describe current local and global activities towards reduction of energy consumption in wireless access networks. Section 12.3 describes our quantitative analysis of the potential for energy saving. Although the analysis is based on simulations of an HSPA network, the achieved conclusions will extend to other access network technologies. This chapter ends with a section summarizing our analysis and conclusions, and a list of references to relevant literature.

12.2 Energy Saving Activities in the Cellular Network Ecosystem

In this section we present an overview of activities undertaken to reduce the energy consumption of cellular networks. As will become clear, all actors in the cellular ecosystem contribute their share to the important aim of energy saving. These actors include academic/research institutes, equipment vendors, network operators, (inter)national governments, industry organizations and standardization forums.

12.2.1 Research Activities

Energy saving for wireless access networks is a hot topic recently in the research/academic community.

The FP7 project SOCRATES[1] [4] sees reduction of energy consumption as one of the use cases for Self-Organising Networks for future wireless networks [5]. Its objective is to reduce OPEX[2] cost and CO_2 emission, and minimize resource usage while ensuring coverage and quality requirements. Energy consumption is reduced by dynamically switching off/on cells, base stations and other radio resources (e.g. transmit antennas), according to observed traffic load, resource utilization, quality and coverage.

---

[1] Self-Optimization and self-ConfiguRATion in wirelEss networkS.

[2] OPerational EXpenditure.

eMobility, a European technology platform for mobile and wireless communications supported by FP6 and FP7, has identified 'green wireless communications' as one of the strategically important technologies for future research in its strategic research agenda [6]. They propose to increase the energy efficiency of wireless communications systems by applying a global optimization approach, considering both equipments and networks. In addition they advocate the use of renewable energy sources.

One common way of energy saving recognized in the literature is dynamically switching off/on cells and/or base stations based on self-organization features that monitor traffic and network conditions (coverage, quality). It is shown in [7] that it is possible to switch off some cells and base stations in urban areas during low-traffic periods, while still guaranteeing quality of service constraints in terms of blocking probability and electromagnetic exposure limits. The reported energy saving is up to 50%. Similar work is reported in [8], which claims energy savings of the order of 25–30%. Marsan and Meo [9] investigated energy saving through energy-aware cooperative management of the cellular access networks of two operators offering service in the same area. Energy is saved by switching off one of the two networks when traffic is so low that the desired quality of service can be obtained with just one network. When one of the networks is off, its subscribers are allowed to roam to the other network which is on. The results show that potential energy saving is in the order of 20%.

Energy-aware signal processing algorithms and network protocols could also help in energy saving. An example can be found in [10], where a control system is proposed to adapt UMTS receiver parameters at run-time to the dynamically changing environment, to reduce energy consumption of mobile terminals while guaranteeing an adequate Quality of Service for the end-user. A methodology is presented in [11] for the design, simulation and optimization of wireless communication networks for maximum performance with an energy constraint. The methodology integrates various aspects of a wireless communication network from the device layer, such as antenna and amplifier design, all the way to the network and application layers. The results illustrate the trade-off between performance and energy consumption in wireless networks.

12.2.2 Vendor Activities

Vendors contribute to the reduction of energy consumption in the design and production period of wireless products (network equipments and terminals).

The main methods can be grouped into three different directions, viz. design and production of energy efficient products, improve energy efficiency of auxiliary devices (e.g. air conditioning, power management, mobile charger, etc.), and use of alternative energy sources.

Advanced signal processing, circuit design, and architecture design are often used by vendors to reduce the power consumption of network equipments, especially base stations which are the highest contributor of CO_2 emissions in wireless networks [12]. One common way is the efficiency improvement of power amplifiers [12, 13], which account for about 50% of a base station's energy consumption. According to Haynes [14] an advanced power amplifier design can improve the efficiency from a typical figure of 15% to more than 50%. Another option for significant improvement of energy efficiency is through the so-called Main Remote solution, also known as tower top-mounted radios [12, 13], or remote radio, where the radio module is installed on a tower near the antenna and thus the conventional feeder loss (about 3 dB) is avoided. Another advantage of the Main Remote solution is that, together with special cabinet design, the main units of the base station could be (partly) cooled by natural convection. Thus the conventional cooling system might be replaced by simple fans or passive heat exchangers, with the exception that the battery compartment is still cooled by a mini air conditioner.

With the reduction of power consumption per base station, the possibility that a base station can be powered by alternative energy sources (e.g. solar/wind power, bio-fuels) becomes higher. This is especially attractive for base stations beyond the reach of the electricity grid, or where the electricity supply is not reliable. Solar power and wind power are two natural energy resources which provide virtually free energy, at least in terms of OPEX [12, 13]. The main investment involved is the additional initial CAPEX[3] for the deployment of solar/wind power equipments, which is coming down as the associated technologies mature. Another advantage of solar/wind power is their low environmental impact. One downside of solar/wind power is the natural variation in sunshine and wind power, introducing a need for sufficient energy storage capacity, or a fall-back power source (e.g. diesel generator).

Most network equipment and terminal vendors have published energy saving targets and solutions, including those in [15–19], covering energy saving in various aspects like products, operation, facilities, etc. Here we

---

[3] CAPital EXpenditure.

illustrate vendors' activities in energy saving by describing the targets and solutions of the leading network equipment and terminal vendors.

Network equipment vendor. Ericsson has released its cooperate responsibility and sustainability report 2008 [15]. It describes the CO_2 emission reductions per GSM or UMTS subscriber since the introduction of new technologies in 1992 and 2001, respectively. Moreover it describes Ericsson's CO_2 emission reduction targets per subscriber for the period 2008–2013. Ericsson's plans cover aspects including raw material extraction, manufacture, transport, use, disassembly and end-of-life treatment of their products [12]. With 2007 as the baseline, Nokia Siemens Networks [16] is committed to improvement of the energy efficiency of its base station products by up to 40% by 2012. Their new Flexi Multiradio base station consumes only 790 W, while running both GSM/EDGE and UMTS/HSPA at the same time. The energy saving methods mentioned include minimizing air-conditioning with parts of the base station located outside and deploying software features to optimize the use of carriers by monitoring the real-time traffic.

Terminal manufacturer. Nokia has published in its environmental strategy [18], among others, energy saving targets for products and services. These include the reduction of the average power consumption of unused phone chargers with 50% between 2006 and 2010. Nokia was the first phone maker to add energy saving alerts to mobile phones, encouraging users to unplug the charger once the battery is full [20] or to disable unused energy-consuming features [21]. Further, Nokia continues to study the use of renewable energy resources, such as solar panels and kinetic energy. Samsung has published its sustainability report 2009 [22], covering among others the goal of product energy efficiency enhancement. It is mentioned in the report that, by the end of 2009, Samsung aims at standby power of 1 W for all its products. Samsung claimed to be the first phone maker to launch solar powered mobile phones [23].

12.2.3 Operator Activities

Since recent years, most leading mobile network operators recognize their responsibility in reducing their CO_2 footprint and/or see substantial monetary benefits in reducing their networks' energy consumption. The strategies generally concentrate on three distinct areas, viz. migrate to the use of renewable energy sources, deployment of energy efficient network equipment and operations, and the development of various products and services for a range of industry sectors that enable energy efficient activities and carbon reductions.

Acknowledging that nowadays most mobile network operators have some form of strategy and defined objectives for contributing to the global need for reducing CO_2 emissions, we choose to illustrate the operator strategies and targets by considering three of the world's largest mobile carriers.

China Mobile has recently signed an agreement to reduce its energy consumption by 20% by the end of December 2012 to save a total of 11.8 billion kWh of electricity over a three-year period [24]. Since about three years all its equipment suppliers must pledge that their products are the most advanced for saving energy. Further, China Mobile thoroughly improved existing equipment with a particular focus on air conditioning systems [25], which are stated to account for a quarter of the company's total electricity consumption. Utilizing solar, wind and bio-fuels as alternative energy sources for over 5000 (of its 400,000) base stations, China Mobile has one of the world's largest deployments of green technologies to power its network [26, 27].

Vodafone states that about 80% of its energy consumption, which in the UK amounts to about 438 GWh per annum, is used in its mobile networks, including radio base stations, switches and routers [28, 29]. It has announced that by 2020 it will reduce its CO_2 emissions by 50% against its 2006/7 baseline of 1.23 million tonnes [29]. This target will be achieved principally by improvements in energy efficiency and increased use of renewable energy. The energy efficiency of its networks is enhanced in different ways, including the use of more energy efficient power amplifiers, remote radio heads, integrated 2G/3G base stations, a reduced need for air conditioning by applying 'free cooling' and upgrading technology to resist higher temperatures, and by switching off base stations when traffic demand is low [30]. As an example of the use of renewable energy, Vodafone claims that about 88% of the energy consumed by its network in the Netherlands is generated by wind turbines [31].

Telefonica has drawn up a 'Manual of good practice for energy efficiency in networks', which sets out more than 50 tried, tested and rapidly implementable methods of reducing/optimising energy consumption, including a review of temperature set points for heating and air conditioning and adjusting power factors [32]. Furthermore, it has defined effective short and long-term strategies to achieve energy efficient integrated networks, regarding energy efficient equipments, planning and management of access and core networks [33].

Verizon Wireless, North America's largest mobile network operator, is also very active in taking 'green initiatives' for reducing its own carbon footprint as well as supporting customers in reducing theirs [34]. Internal measures include the use of solar cell sites in order to power the wireless network, enhancing the energy efficiency of the retail stores and utilising energy-saving technology on company desktops and workstations. In an effort to cut back on its power consumption and energy costs and shrink its carbon footprint, mother company Verizon has developed its own energy-consumption standards, accompanied by a measurement process for telecommunications equipment [35–37]. The purpose of these standards is 'to foster the creation of more energy efficient telecommunications equipment by Verizon's supplier community thereby reducing the energy requirements in Verizon networks' [36].

In a recent report [38] Vodafone, in collaboration with Accenture, demonstrates that mobile technology can be a major catalyst in driving carbon reductions across a range of industry sectors, potentially cutting Europe's annual energy bill by at least €43 billion. Among the identified opportunities are the applications in smart electricity grids, smart logistics and smart manufacturing. Similar initiatives are taken by other network operators, including China Mobile, Telefonica, and Verizon [26, 34, 39, 40].

12.2.4 Industry Organizations

In this section we address the activities of two key organizations in the wireless industry, involving primarily network operators and equipment vendors.

The NGMN (Next Generation Mobile Networks) [45] alliance and its partners place a strong emphasis on the development of the most energy efficient next generation technology and on the extension of the industry's green footprint. Therefore, the NGMN Green Telco Initiative brings together NGMN partners who run projects and have started activities to protect the environment [46]. In addition, NGMN has identified energy reduction as one of their 'self organization use cases' [47].

The GSMA and The Climate Group have unveiled their 'Mobile Green Manifesto' [1], outlining four goals to reduce the mobile industry's greenhouse gas emissions: (1) to reduce the total global greenhouse gas emissions per connection by 40% by 2020 compared to 2009; (2) to ensure carbon neutral growth of mobile networks, with the number of mobile connections set to rise by 70% until 2020; (3) to reduce the energy consumed by a typical handset, both in standby and active modes by 40% by 2020; and (4) to reduce the life cycle emissions of network equipment components by 40% by 2020.

With its 'Green Power for Mobile' program, the GSMA is aiding the mobile industry to deploy solar, wind, or sustainable bio-fuels technologies to 118,000 new and existing off-grid base stations in developing countries by 2012 [49]. The GSMA stated that network optimization upgrades (through advanced planning and spectrum allocation) currently can reduce energy consumption by 44% and solar-powered base stations could reduce carbon emissions by 80% [50]. The GSMA is also working with seventeen leading mobile operators and manufacturers in implementing a Universal Charging Solution (UCS) for mobile phone chargers, which is expected to roll out in worldwide markets by 2012 [51]. This initiative aims to ensure that the mobile industry adopts a common format with higher energy efficiency rating for mobile phone chargers. The UCS-based chargers are expected to be up to three times more energy efficient than conventional chargers. It is expected that, with the UCS solution, the number of chargers being manufactured reduces annually by 50% and the associated greenhouse emission reduces by 13.6 million to 21.8 million tonnes per year. From 2009, the GSMA sets special 'Green Mobile Awards' for the mobile industry [52], highlighting the role that mobile can play in the development of low carbon economies, industries and lifestyles.

12.2.5 3GPP Standardization

Also in standardization, the subject of energy saving and energy efficient wireless communication is an important one. Historically, this has always been an important issue from the mobile terminal point of view due to the limited battery power resources. For example, in wireless access systems the mobile terminal activity in idle mode (i.e. when there is no active session between the terminal and the network) is reduced to the bare minimum in order to save battery power. In this *discontinuous reception* (DRX) mode, the transceiver electronics of the mobile is switched off and is temporarily waked up to listen to the downlink signalling channels in order to e.g. perform mobility related procedures and identify whether there is incoming call (paging). In newer systems like LTE and WiMAX, DRX is also introduced in connected mode. Another mechanism that reduces energy consumption is *transmit power control*, which adjusts transmission powers in order to optimize network capacity and service quality. There are also other indirect examples of energy efficient transmission/reception in active mode such as the so-called *continuous packet connectivity* in [53], which is focused on reducing overhead and resource consumption for 'always-on' like wireless sessions or VoIP sessions but indirectly also prolonging the battery life of the mobile terminal.

In recent years standardization bodies such as 3GPP have acknowledged the goal of energy saving in a broader sense. In the 3GPP working groups RAN3 and SA5 standardization of energy saving features has been started for UMTS/HSPA and LTE networks via switching on/off of sectors and base stations. In RAN3 a discussion is ongoing for energy saving actions in UMTS/HSPA [54] where the RNC makes decisions about switching on/off the base stations and signals the decisions and corresponding parameters towards the base stations. In SA5 a study on energy saving management for LTE/SAE wireless access has been started [55]. Several scenarios are defined for investigation of the energy saving actions; architectural options are defined, etc. These RAN3 and SA5 activities are focused on energy efficient operation of the wireless access network where the amount of active/switched-on base stations is dynamically adjusted according to the traffic demands and coverage/quality targets as defined by the wireless operator.

An alternative dimension of energy saving in wireless access network is to make the working of the base stations as energy efficient as possible. The 3GPP standardization group RAN1 has started a recent study item [56] with the objective to identify potential solutions to enable energy saving within UMTS base stations, considering its impact on service accessibility, power consumption and mobility performance.

These activities should propose solutions for energy efficient UMTS/HSPA and LTE networks. In the future it can be expected that 3GPP will investigate energy saving features for LTE Advanced and also for femtocell deployments.

12.2.6 Governments

With increasing environmental and energy challenges facing the world, governments and governmental organizations spend more and more attention on the relationship between ICT and the environment. This relationship mainly consists of two aspects, i.e. how ICT can enable the energy reduction in the whole economy and how ICT reduces the energy consumption by itself. Most governments and governmental organizations focus their policies and analysis on the ICT industry in general, rather than concentrating specifically on the mobile/wireless industry.

The Organization for Economic Co-operation and Development (OECD) is developing policies, undertaking analysis, and facilitating international debate, on the use of ICT to tackle environmental challenges. The OECD has published a report on measuring the relationship between ICT and the environment [41], which introduces a conceptual framework for statistically analyzing the impact of ICT on the environment.

The European Commission (EC) sees a significant enabling role for ICT to achieve its political aim of energy reduction across the whole economy [42]. At the same time, the Commission made clear that the ICT industry should 'lead by example' in the drive towards CO_2 reductions, significantly cutting its own emissions. More recently, the EC wanted the ICT sector to 'show the way for the rest of the economy by already reducing its own carbon footprint by 20% by 2015' [43]. As an example of an energy reduction policy at the national level, the Dutch government has signed an agreement with the ICT sector to improve its energy efficiency with 30% in the period 2005–2020.

The US government also recognizes the important role of ICT in the energy efficiency improvement of the American economy. It was announced that a funding of $50 Million will be invested to conduct research, development, and demonstration projects to promote new technologies that improve energy efficiency in the ICT sector [44].

12.3 Energy Saving Potential of Switching Off Base Stations in Off-Peak Hours

Mobile communication networks are usually planned to provide some minimum service quality level during peak traffic hours. Consequently, in off-peak hours, when traffic loads are lower, the network is characterized by over-capacity, in the sense that same service quality targets can typically be satisfied with a reduced set of network resources, e.g. sites, carriers, etc. In this section, we propose a procedure for deriving the potential of energy-oriented network optimization, and apply this procedure to the case of a UMTS/HSDPA network.

The outline of this section is as follows. Section 12.3.1 presents the overall approach in analysing the potential of energy-oriented network optimization. Section 12.3.2 describes the typical UMTS/HSDPA network scenario model that we use for application of the optimization approach. In Section 12.3.3 we analyse the capacity of an extensive set of reduced network configurations, given pre-specified performance requirements. Section 12.3.4 presents the principle and gains of hour-by-hour energy-optimized network configuration. The section is finalized with a sensitivity analysis for the energy savings with respect to the energy model parameters and the applied performance requirement in Section 12.3.5.

12.3.1 Approach

Starting point of the analysis is an UMTS/HSDPA network of 48 sites with three sectors, placed in a hexagonal layout with an applied wraparound technique to mimic an infinite network and avoid boundary effects as presented in Figure 12.1. For each site the sectors are equipped with four 5 MHz carriers. The inter-site distance and antenna downtilt are optimized to handle the peak hour traffic load with a pre-specified quality of service and coverage requirements, assuming full availability of all sites, sectors and carriers. The following three-step procedure is the selected approach for quantifying the energy consumpti on gain:

Step 1 – We derive the relation between the considered performance indicators and the traffic load, for an extensive set of 'reduced network configurations'. These 'reduced network configurations' are specified by the set of active sites, starting from 48 down to 12 sites as presented in Figure 12.1, and the uniform number of active carriers per sector. The applied performance indicators are the average flow throughput and the 10th flow throughput percentile. From these results we derive the maximum supportable traffic load for each network configuration. This is presented in Section 12.3.3.

Step 2 – Considering a typical pattern of daily traffic load fluctuations, we determine for each hour of the day whether the different configurations (from Step 1) are indeed able to support the traffic load with adequate service quality, and if so, what the induced network-wide energy consumption level is. For these energy calculations, an energy consumption model is applied, which depends on the network configuration and the average transmit powers. Among those network configurations able to support the load, the one that minimizes energy consumption is selected as 'optimal'. This is presented in Section 12.3.4.

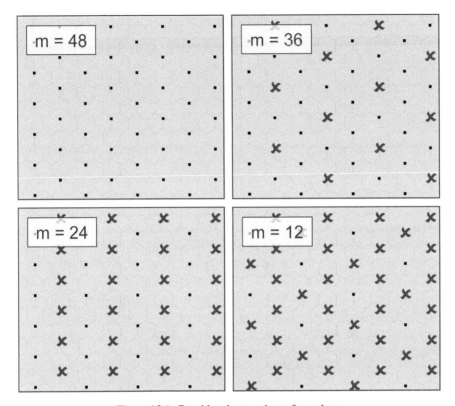

Figure 12.1 Considered network configurations.

Step 3 – Comparing the daily energy consumption of the per-hour energy-optimized network configuration with a 'full configuration' provides us with an indication of the energy savings potential when switching off over-capacity in off-peak hours. This is presented in Section 12.3.4.

It is noted that several aspects of the conducted study are debatable, in the sense that they reflect just one of several sensible choices that can be made in the analysis, e.g. the selected daily traffic pattern, the choice for a hexagonal network layout, the considered propagation environment or the assumed energy consumption model. Besides the numerical outcome in the form of an indication of the energy savings potential, which in our view is based on an appropriate set of modelling assumptions, we believe that the

Table 12.1 Propagation/interference model.

Path loss	$123.22 + 35.22 \log_{10} d_{km}$
Antenna diagram	Kathrein 741989, 17 dBi gain
Feeder/slant/penetration loss	13.5 dB
Shadowing	$\sigma = 6$ dB, inter-site correlation $= 1/2$
Orthogonality factor	0.10
Noise figure	8 dB

general approach to derive this outcome is another valuable contribution of the presented work.

12.3.2 Reference Model

The reference model considers $m = \{12, 24, 36, 48\}$ active sites with three sectors, where each sector is equipped with $n = \{1, 2, 3, 4\}$ carriers. Each carrier is characterized by a maximum transmit power of 20 Watt, of which 2 Watt is assigned to the CPICH.[4] The remaining 18 Watt is assigned to the HS-DSCH.[5] The HS-DSCH features a round robin packet scheduler and applies a Shannon-based link adaptation scheme to map the experienced SINR[6] to an applied bit rate (in Mb/s):

$$r(\text{SINR}) = \min\left\{n \times 21.6, n \times 5 \times \log_2(1 + \text{SINR})\right\},$$

where n is the number of active carriers and 21.6 Mb/s is the per carrier peak rate. The specifics of the applied *propagation/interference model* are summarized in Table 12.1.

Considering the selected propagation model, the inter-site distance and the antenna downtilt have been to 1.1 km and 4°, respectively. These settings are made in correspondence with the assumed peak hour traffic load, while it has been further verified that even in a reduced network configuration with twelve active sites (and unaffected downtilt), sufficient uplink (voice service, assuming a 6 dB noise rise) and downlink (CPICH) coverage exist. This consequently means that basic coverage (e.g. emergency voice calls) is always provided.

The *traffic model* is characterized by the consideration of a downlink data service. The traffic load, expressed in terms of the network-wide average number of active data flows, varies during the day according to the profile

---

[4] Common PIlot CHannel.

[5] High Speed Downlink Shared CHannel.

[6] Signal to Interference plus Noise Ratio

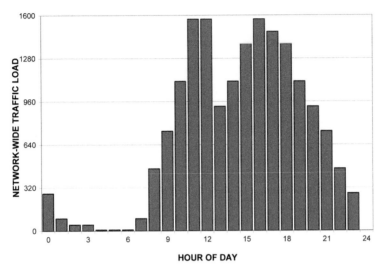

Figure 12.2 Daily traffic load variations.

depicted in Figure 12.2. While the basic (relative) traffic load profile is taken from actual network measurements, it has been scaled to match an assumed peak hour traffic load of about 3466 or 1508 data flows, depending on whether we consider a network operator that applies an average throughput requirement of 1 Mb/s or a requirement of 400 kb/s for the 10th throughput percentile (see also below). This peak hour traffic load is set in correspondence to the considered system model and the pre-specified performance targets (see also Section 12.3.3).

12.3.3 Analysis of Reduced Network Configurations

We make use of Monte Carlo simulations to derive the relation between the experienced performance and the offered traffic load. This is done for an extensive set of network configurations, which are specified by the number m of active sites and the number n of active carriers per sector. Herein, m is taken from $\{12, 24, 36, 48\}$ and n is taken from $\{1, 2, 3, 4\}$, constituting sixteen distinct configurations.

Figure 12.1 shows which of the 48 sites are switched off in the reduced configurations where $m = 12, 24$ or 36.

The results of these numerical experiments are presented in Figure 12.3. The charts show the average flow throughput, the 10th flow throughput percentile and the activity factor as a function of the network-wide traffic load.

The activity factor ρ is defined as the fraction of time that a sector's HS-DSCH is transmitting, i.e. the fraction of the number of snapshots where a sector serves at least one data flow.

This activity factor is readily converted to a power activity factor γ, viz.

$$\gamma = \frac{(1 - \rho) \cdot 2 + \rho \cdot 20}{20} = 0.1 + 0.9\rho,$$

which impacts the energy consumption, as we will see below.

All qualitative results presented in Figure 12.3 are as intuitively expected: the 10th throughput percentile is always lower than the average throughput, and both throughput metrics are decreasing in the traffic load and increasing in the number of active sites and the number of active carriers. The activity factor is also nicely increasing in the traffic load, independently of the number of active carriers.

Assuming a target level of 1 Mb/s for the average throughput or 400 kb/s for the 10th throughput percentile, the maximum supported traffic load is determined for each network configuration (m, n). These results are presented in Figure 12.4.

As already briefly mentioned in Section 12.3.2, it is based on these results that we have set the peak hour traffic load to a network-wide average of about 3466 (1508) active data flows for the case with an average throughput (10th throughput percentile) requirement of 1 Mb/s (400 kb/s), i.e. sensibly assuming that the fully configured network is appropriately dimensioned for the offered traffic load in peak hour. The fact that the supportable traffic loads are significantly higher for case with an average throughput requirement indicates that the target of 1 Mb/s is a relatively weak requirement compared to the 400 kb/s requirement for the 10th throughput percentile that is applied to obtain the chart on the right.

Observe further from these results that in distinct scenarios with the same number of deployed sector-carriers (the number of sectors × the number of carriers), it is not trivial which configuration supports the highest load. Consider for instance the right chart that is based on the 10th throughput percentile requirement. If we compare a configuration with 12 active sites and 3 carriers per sector (108 sector-carriers) with a configuration with 36 active sites and 1 carriers per sector (also 108 sector-carriers), the former configuration supports the highest traffic load, i.e. rather more carriers than more sites. On the other hand, if we compare a configuration with 24 active

12 ACTIVE SITES

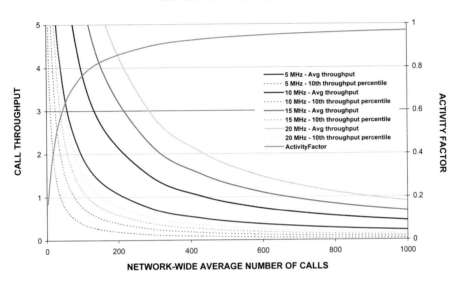

24 ACTIVE SITES

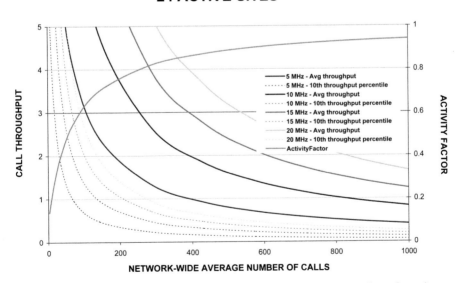

Figure 12.3 Throughput performance versus traffic load for diverse network configurations.

36 ACTIVE SITES

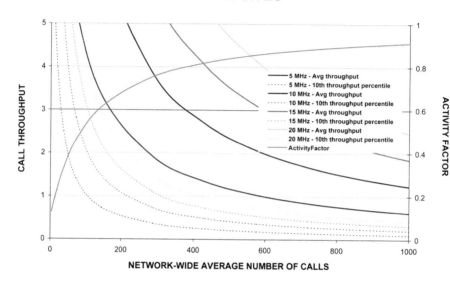

48 ACTIVE SITES

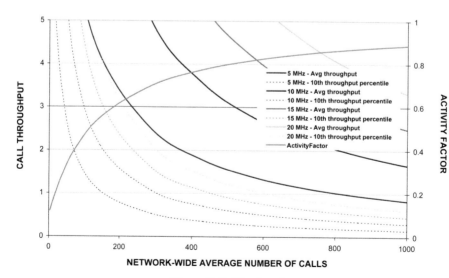

Figure 12.3 (Continued)

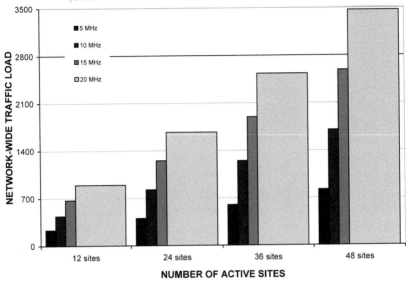

Figure 12.4 Maximum supportable network/wide traffic load per network configuration. The left (right) chart corresponds with the case where a 1 Mb/s (400 kb/s) requirement is applied for the average throughput (10th throughput percentile).

sites and 3 carriers per sector (216 sector-carriers) with a configuration with 36 active sites and 2 carriers per sector (also 216 sector-carriers), it is the latter configuration that supports the highest traffic load, i.e. rather more sites than more carriers.

12.3.4 Energy-Oriented Network Optimization

The next step in the analysis is to determine for each hour of the day (see Figure 12.2) the most appropriate network configuration, i.e. the most energy efficient network configuration that can handle the traffic load.

For this purpose it is necessary to introduce an *energy consumption model*, which translates the number of active sites, the number of active carriers (in an active sector) and the activity factor to an aggregate (network-wide) energy consumption level. The energy consumption model comprises two main components: the first component captures the energy consumption of the air conditioning unit, while the second component captures the energy consumption of the baseband/radio frequency (BB/RF) unit.

Denote with S_{on} the number of active sites and with $S_{off} = 48 - S_{on}$ the number of inactive sites. Further, denote with C_{on} the number of active carriers (in an active sector) and with $C_{off} = 4 - C_{on}$ the number of inactive carriers (in an active sector). Recall that ρ denotes the activity factor. The applied energy consumption model is characterized by four parameters:

- P_{AC} is the energy consumption of a fully active air conditioning unit per site;
- $P_{BB/RF}$ is the energy consumption related to the BB/RF unit per fully active carrier; this captures all energy consumption not related to the air conditioning unit;
- α is the fraction of P_{AC} that is the minimum energy consumption of an air conditioning unit, regardless of the site-sector's (in)activity;
- β is the fraction of $P_{BB/RF}$ that is the minimum energy consumption of the BB/RF unit per carrier, regardless of the sector/carrier's (in)activity.

Applying a linearity assumption of energy consumption w.r.t. sector activity, the following formula is used to determine the energy consumption:

$$E = S_{off} \left(\alpha P_{AC} + 4\beta P_{BB/RF} \right)$$
$$+ S_{on} \left(\begin{array}{l} P_{AC} \max \left\{ \alpha, \tfrac{1}{4} C_{on} \gamma \right\} \\ + C_{on} P_{BB/RF} \max \left\{ \beta, \gamma \right\} + C_{off} \beta P_{BB/RF} \end{array} \right),$$

Here, the first part, $S_{off}(\alpha P_{AC} + 4\beta P_{BB/RF})$, covers the energy consumption of inactive sites. The second part covers the energy consumption of the active sites, where:

- $P_{AC} \max\{\alpha, \frac{1}{4}C_{on}\gamma\}$ covers the energy consumption of the AC unit depending on carrier activity;
- $C_{on} P_{BB/RF} \max\{\beta, \gamma\}$ covers the energy consumption of the BB/RF unit depending on carrier activity;
- $C_{off}\beta P_{BB/RF}$ covers the energy consumption of the inactive carriers at the active site.

We assume $P_{AC} = 1500$ Watt and $P_{BB/RF} = 750$ Watt, consider different choices of α and β, and note that the power activity factor γ is readily obtained from the simulations (see also Figure 12.3). For the case of $\alpha = \beta = 10\%$ the results in Figure 12.5 show for each hour of the day the energy consumption level if we constantly deploy a configuration with 48 sites and 4 carriers per sector, as well as the energy consumption level if in each hour of the day the most energy efficient configuration is deployed (which still satisfies the imposed quality of service target). Additionally, for the latter case the figure indicates the optimal configuration S_{on}/C_{on}, written inside the bars. For the case with an average throughput requirement of 1 Mb/s, the average network-wide energy consumption level is about 0.167 MW for the default scenario and 0.113 MW for the energy-optimal scenario, which constitutes a savings of almost 32%. For the alternate case with a 10th throughput percentile requirement of 400 kb/s, the average network-wide energy consumption level is about 0.148 MW for the default scenario and 0.106 MW for the energy-optimal scenario, which constitutes a savings of almost 29%.

12.3.5 Sensitivity Analysis

In this section, we assess the sensitivity of the energy savings w.r.t. the choices of α, β, as these parameters may take a range of settings in practical implementations. Figure 12.6 shows the results. Observe that the energy savings are largest for the case with a planning target on average user throughput. As could be expected, the energy savings are decreasing in both α and β. For the considered range of settings the energy savings vary between about 18 and 38%, also depending on the applied quality of service requirement.

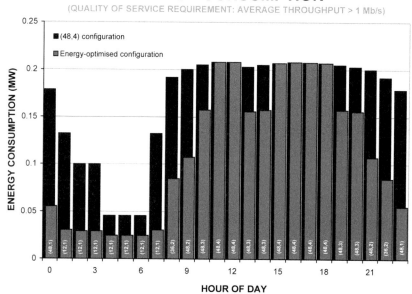

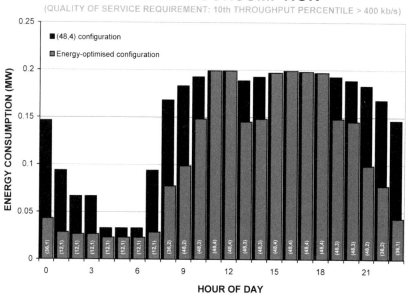

Figure 12.5 Default and optimized energy consumption levels for different hours of the day.

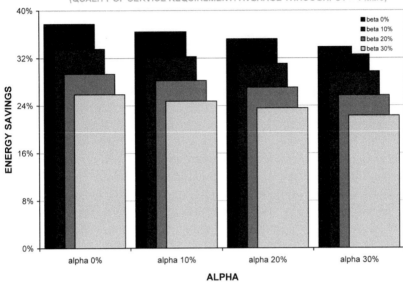

Figure 12.6 Sensitivity assessment of the energy savings w.r.t. energy model parameters α and β.

12.4 Summary and Recommendations

The worldwide goal of mitigating the negative climate changes via CO_2 emissions reduction and the need for energy consumption reduction as part of the OPEX reduction have strongly influenced the ICT sector. The ICT systems are currently contributing with 2 to 3% in the global energy consumption (or CO_2 emissions) with a tendency to increase in the future. Although the ICT systems can be deployed to reduce energy consumption or to enable power generation they can also operate in more energy efficient way. The focus of this chapter is energy efficiency in the area of wireless access networking.

It can be concluded that energy efficient wireless access networks are widely supported by equipment/terminal vendors, network operators, (inter)national governments and standardization bodies. Network and terminal vendors are developing new concepts for more energy efficient and CO_2 friendly equipment and manufacturing processes. Network operators, besides acquiring energy efficient equipment, further reduce their CO_2 emissions by advanced mechanisms for switching on/off network resources and by using renewable energy sources. 3GPP is also facilitating energy efficiency by starting working items and specification activities on energy saving for UMTS and LTE. This wide support of the key players in the wireless access ecosystem increases the probability of achieving the ambitious CO_2 emission targets.

One promising approach for energy saving in wireless access networks presented in this chapter is the dynamic switch on/off of base stations or sectors as the traffic demand varies during the day. We have presented a procedure for deriving the potential energy savings from this approach and have applied this to a UMTS/HSDPA network. The achievable energy savings are in the range of 18 to 38%, depending on the network operator's performance target and the specifics of the energy consumption model.

Our recommendations for further work on energy saving/efficiency in wireless access networks are twofold.

First, quantitative studies on energy saving via base stations (or sector) switch on/off can be improved with enhanced and verified energy consumption models, which also incorporate temperature dependencies, and extended by considering multi-RAT and multi-band scenarios with deployed GSM/EDGE, UMTS/HSPA and/or LTE wireless access networks, considering advanced LTE concepts with relaying and repeaters, and evaluating the effects of femtocells on the energy consumption. Furthermore, there is a need to develop centralized and/or distributed algorithms that can adaptively switch on and off network resources in response to monitored traffic load and/or experienced service quality levels. Consequently, the applicable technology standards may need to be adapted to provide the means for such adaptive, energy-oriented network optimization algorithms.

Secondly, the energy saving research in the ICT field aims at achieving energy *efficiency* targets. This introduces the need to define, measure and benchmark energy efficiency of (mobile) telecommunication networks. This is a challenging, but not a new task. For example, Telecom Italia has deployed (since 2003) a system to monitor the energy efficiency based on 'joule per bit', taking the combined energy usage of all energy consuming elements of the company (network infrastructure, buildings, vehicle fuel, etc.) and representing all traffic (including voice) as an equivalent number of data bits. Although this is a valuable approach, it does not provide means to compare energy efficiency of different network services provided by different network operators. It is generally not straightforward to split the energy bills over different services, and moreover, energy bills will often be applicable to different parties. For instance, it is not straightforward how to split the energy consumption of a shared base station site over the involved operators. In addition, the service quality will have a strong effect on energy usage. E.g. a mobile network with worse coverage would typically use less energy than one with perfect coverage, most likely due to a denser deployment of base stations. To realize a fair 'apples-to-apples comparison, these effects should somehow be incorporated in the energy efficiency indicator. For the time being, the definition of a fair, useful and implementable energy efficiency indicator for wireless access networks is an open research problem. In the quantitative analysis included in this chapter we circumvent this open problem, by looking at the reduction of energy usage of a *given* network, assuming a *given* daily traffic profile and comparing *distinct* service quality targets.

Acknowledgements

We acknowledge the contribution of and fruitful discussions with our TNO colleagues Nicholas Chevrollier, Rob Mulder, Marijn Rijken and Adrian Pais.

References

[1] GSMA, Mobile Green Manifesto, Available: http://www.gsmworld.com/documents/mobiles_green_manifesto_11_09.pdf.

[2] GSMA, www.gsmworld.com.

[3] 3GPP, http://www.3gpp.org.

[4] SOCRATES, http://www.fp7-socrates.eu.

[5] SOCRATES Deliverable D2.1. Use Cases for Self-Organizing networks, March 2008, Available: http://www.fp7-socrates.eu.

[6] eMobility, Strategic research agenda, Available: http://www.emobility.eu.org/SRA/Documents/eMobility_SRA_07_090115.pdf.

[7] L. Chiaraviglio et al., Energy aware UMTS access networks. In *Proceedings of the 11th International Symposium on Wireless Personal Multimedia Communication WPMC 08*, Lapland, Finland, September 2008.

[8] M.A. Marsan, et al., Optimal energy savings in cellular access networks. In *GreenComm 2009*, Dresden, Germany, June 2009.

[9] M.A. Marsan and M. Meo, Energy efficient management of two cellular access networks. In *SIGMETRICS/Performance 2009*, Seattle, WA, USA, June 2009.

[10] L.T. Smit et al., Energy-efficient wireless communication for mobile multimedia terminals, Available: http://eprints.eemcs.utwente.nl/1482/01/smit1_radiomatics04.pdf.

[11] W. Stark, et al., Low energy wireless communication network design, Available: http://www.eecs.umich.edu/aromuri96/STA00I.pdf.

[12] Ericsson White Paper, Sustainable energy use in mobile communications, http://www.ericsson.com/campaign/sustainable_mobile_communications/downloads/sustainable_energy.pdf.

[13] Huawei, The green CDMA base station, Available: http://www.huawei.com/publications/view.do?id=5715\&cid=10549\&pid=61.

[14] Haynes, T., Designing energy-smart 3G base stations, *Mobile Dev & Design*. Available: http://mobiledevdesign.com/hardware_news/radio_designing_energysmart_base/, 2007.

[15] Ericsson, Ericsson corporate responsibility and sustainability report 2008, Available: http://www.ericsson.com/ericsson/corporate_responsibility/cr08_doc/corporate_responsibility_%20report_2008.pdf.

[16] Nokia Siemens Networks, Carbon cuts through enhanced technology, Available: http://www.nokiasiemensnetworks.com/sites/default/files/WWF_Climate_Savers.pdf.

[17] Alcatel Lucent, Cooperate social responsibility report 2008, Available: http://www.alcatel-lucent.com/csr/csr-report/Alcatel-Lucent-CSR-Report-2008-EN.pdf.

[18] Nokia, Environmental strategy, Available: http://www.nokia.com/environment/our-responsibility/environmental-strategy/energy-saving-targets.

[19] Huawei, Energy saving & CO_2 emission cut at Huawei, Available: http://www.huawei.com/publications/view.do?id=2884\&cid=5269\&pid=127.

[20] Nokia, Nokia becomes the first maker to add energy saving alerts to mobiles, http://www.nokia.com/press/press-releases/showpressrelease?newsid=1125979.

[21] Nokia, Green your Nokia cell phone with energy-saving GreenPhone App, Available: http://www.riverwired.com/blog/green-your-nokia-cell-phone-energy-saving-greenphone-app.

[22] Samsung, Sustainability report 2009, Available: `http://www.samsung.com/us/aboutsamsung/sustainability/sustainabilityreports/download/2009/2009Environmentalnsocialreport.pdf`.

[23] Rethink Wireless, Samsung and ZTE claim solar handset firsts, Available: `http://www.rethink-wireless.com/article.asp?article_id=1043`.

[24] China CSR, China mobile commits to energy conservation, see `http://www.chinacsr.com/en/2009/11/16/6576-china-mobile-commits-to-energy-conservation`, November 2009.

[25] Nokia Siemens Networks, China Mobile on course to hit 40 percent energy reduction target by 2010, see `http://www.nokiasiemensnetworks.com/sites/default/files/document/Yunnan_MCC_success_story_Green_project.pdf`, 2009.

[26] China Daily, China Mobile makes an ambitious bid to save energy, see `http://www.chinadaily.com.cn/bizchina/2009-06/22/content_8964637.htm`, June 2009.

[27] GSMA, Renewable energy powered base stations, China Mobile, China, `http://www.gsmworld.com/our-work/mobile_planet/mobile_environment/4252.htm`, 2009.

[28] Simon Mendham (Vodafone), The implementation of AMR metering at Vodafone UK, see `http://www.esta.org.uk/EVENTS/2008_02_19_aMT%20Conference/documents/ESTAaMT20082.4VodafoneMendhamweb.pdf`, February 2008.

[29] Vodafone, Vodafone group announces commitment to reduce CO_2 emissions by 50%, `http://www.vodafone.com`, press release, April 2008.

[30] Vodafone, Climate change, see `http://www.vodafone.com/start/responsibility/environment/climate_change/improving_network.html`, 2009.

[31] Vodafone, Vodafone zet antennes op standby en helpt het milieu, `http://www.vodafone.nl`, press release, December 2008.

[32] Telefonica, Energy efficiency, `http://www.telefonica.com/ext/rc08/en/telefonica/ESPECIAL_M_AMBIENTE/Cambio_Climatico/Eficiencia_energetica/index.html`, 2009.

[33] Marisa Vargas (Telefonica), Energy efficiency & Green IT, `http://www.ogf.org/OGF25/materials/1530/Energy+efficiency+in+the+Telecom+sector+-+Marisa+Vargas+-+Telefonica.pdf`, Open Grid Forum, March 2009.

[34] Verizon Wireless, Verizon Wireless green initiatives, see `http://aboutus.vzw.com/Green_Initiative/Green%20Initiatives%20Highlights.pdf`, 2009.

[35] Verizon, Green technology – Verizon completes energy efficiency standards for network, see `http://green.tmcnet.com/topics/green/articles/30307-verizon-completes-energy-efficiency-standards-network.htm`, 2008.

[36] Verizon, Verizon NEBSTM compliance: Energy efficiency requirements for telecommunications equipment, `http://www.verizonnebs.com/TPRs/VZ-TPR-9205.pdf`, 2009.

[37] Verizon, Verizon NEBSTM compliance: TEEER metric quantification, `http://www.verizonnebs.com/TPRs/VZ-TPR-9207.pdf`, 2009.

[38] Vodafone/Accenture, Carbon connections: Quantifying mobiles role in tackling climate change, see `http://www.vodafone.com/etc/medialib/cr_09/carbon.Par.76396.File.tmp/carbon_web_2009.pdf`, July 2009.

[39] Telefonica, Efficient products and services, `http://www.telefonica.com/ext/rc08/en/telefonica/ESPECIAL_M_AMBIENTE/Cambio_Climatico/Productos_eficientes/index.html`, 2009.

[40] Verizon, `http://newscenter.verizon.com/kit/green-press-kit/`, 2009.

[41] Organization for Economic Co-operation and Development (OECD), Measuring the relationship between ICT and the environment, Available: `http://www.oecd.org/dataoecd/32/50/43539507.pdf`.

[42] European Commission, Addressing the challenge of energy efficiency through Information and Communication Technologies, Available: `http://ec.europa.eu/information_society/activities/sustainable_growth/docs/com_2008_241_1_en.pdf`.

[43] European Commission press release, Commission pushes ICT use for a greener Europe, Available: `http://europa.eu/rapid/pressReleasesAction.do?reference=IP/09/393\&format=HTML\&aged=0\&language=EN\&guiLanguage=en`.

[44] US Department of Energy, Secretary Chu announces $256 million investment to improve the energy efficiency of the American economy, Available: `http://www.energy.gov/news2009/print2009/7434.htm`.

[45] NGMN Alliance, `http://www.ngmn.org`.

[46] NGMN Alliance, NGMN green telco initiative, Available: `http://www.ngmn.org/nc/workprogramme/greeninitiave.html`.

[47] NGMN Alliance, NGMN Informative List of SON Use Cases, Available: `http://www.ngmn.org/fileadmin/user_upload/Downloads/Technical/NGMN_Informative_List_of_SON_Use_Cases.pdf`.

[48] The Climate Group, `http://www.theclimategroup.org`.

[49] GSMA, Green power for mobile, Available: `http://www.gsmworld.com/our-work/mobile_planet/green_power_for_mobile/index.htm`.

[50] GSMA, Mobile and the environment: Mobile networks, Available: `http://www.gsmworld.com/our-work/mobile_planet/mobile_environment/mobile_networks.htm`.

[51] News release, Mobile industry to launch energy efficient universal charger by 2012, `http://www.environmentalleader.com/2009/02/18/mobile-industry-to-launch-energy-efficient-universal-charger-by-2012/`.

[52] GSMA, Global mobile awards 2010: Green mobile awards, `http://www.globalmobileawards.com/cat/cat5.shtml`.

[53] 3GPP TR 25.903, Continuous connectivity for packet data users (Release 9), v9.0.0, December 2009.

[54] 3GPP R3-092342, Energy saving in UTRAN, 3GPP TSG-RAN WG3#65bis, Miyazaki, Japan, 12–15 October 2009.

[55] 3GPP TR 32.826, Study on Energy Savings Management (ESM) (Release 9), v1.0.0, December 2009.

[56] 3GPP RP-091439, Solutions for energy saving within UTRA Node B, 3GPP TSG-RAN Plenary #46, Sanya, China, 1–10 December 2009.

13

Optimizing Energy Usage in Private Households

John Rohde, Sune Wolff, Thomas Skjødeberg Toftegaard, Peter Gorm Larsen, Kenneth Lausdahl, Augusto Ribeiro and Poul Ejnar Rovsing

Aarhus School of Engineering, Aarhus University, Ny Munkegade, Building 1521, 8000 Aarhus C, Denmark

Abstract

The exploding power consumption in private households is a growing challenge in the pursuit of reduction of the worldwide emission of green house gases (GHG). Development of smart sensors and advanced control systems has made it possible to make technology assist in solving this challenge. Complexity and magnitude of the challenge is presented in this chapter and the Minimum Configuration Home Automation (MCHA) framework enabling interoperability between existing communication protocols is offered as an intelligent part of the solution. An outline of technology trends supporting the promotion of the use of ICT in solving the challenge is also included to emphasize the vast potential in the application of ICT in the solutions for the future.

Keywords: Energy management, communications infrastructure, home automation, sensor network, protocol interoperability.

13.1 Introduction

Growing energy demand and the outlook to an increasingly more complex and decentralized energy production emphasizes the need for intelligent and

R. Prasad et al. (Eds.), Towards Green ICT, 185–209.

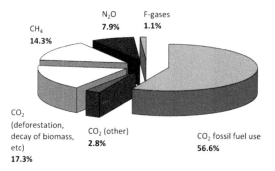

Figure 13.1 Anthropogenic GHG: total emissions (2004) (CO_2-equivalents) [13].

efficient energy consumption management systems in private households in order to bring down the green house gas (GHG) emissions from this important sector.

According to the International Energy Agency (IEA) World Energy Outlook 2008 the world primary energy demand will grow 1.6% per year the next 20 years [19]. Today, a total of more than 12 Mtoe is consumed and more than 80% of the energy originates from fossil fuels. The use of fossil fuels makes a rather substantial contribution to the world emission of CO_2. The relative anthropogenic green house gas contribution in 2004 is shown in Figure 13.1.

Looking at how the different sectors contribute to the total anthropogenic GHG emissions leads to the data for 2004 presented in Table 13.1. The contribution from residential and commercial buildings is 7.9% of the total emissions. The data reflects the fact that this sector is by no means one of the major contributors. Nevertheless, the sector is by far one of the most interesting when it comes to the estimated economical mitigation potentials with respect to GHG reduction in the 2030 forecast done by IEA. When it comes to potential weighted by economy it outperforms all other sectors including industry, transportation and agriculture.

Implementation of efficient energy management in residential buildings is a way to release a large portion of the stated potential reduction of CO_2 emission in the sector. The core of energy management is communications infrastructure supporting highly intelligent sensors and control units. The challenge is to implement flexible, easy-to-install and easy-to-use solutions with built-in intelligence capable of i.e. timing the energy consumption with the availability of surplus energy from renewable energy sources. Cost is of course a major parameter when making solutions for residential buildings and

Table 13.1 Anthropogenic GHG: sector emissions (2004) (CO_2-equivalents) [13].

Sector	Rel. GHG emissions
Residential and commercial buildings	7.9%
Industry	19.4%
Agriculture	13.5%
Forestry	17.4%
Waste and wastewater	2.8%
Energy supply	25.9%
Transport	13.1%

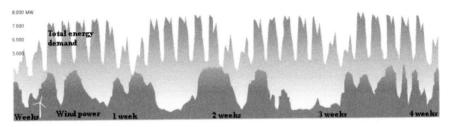

Figure 13.2 The fluctuating production of wind power does not match the consumers demands.

this disqualifies todays available building management system solutions for commercial buildings as an option. A holistic approach is necessary in order to break down the barriers – both economical, human and technological – existing today and preventing full-scale penetration of energy management in residential buildings.

With respect to energy production, increasing the production of renewable energy is seen as one of the main steps towards lowering GHG emissions. The consequence of this is increased complexity and decentralization of the energy production resulting in severe challenges in maintaining and expanding the energy production infrastructure. In the nordic countries (Denmark, Sweden, Norway, Iceland and Finland) the planned increase in production of windpower is 2740 MW by 2012 [8], and Germany has planned to increase the production of windpower by 9000 MW by 2020. The fact that the energy production from renewable energy sources like windpower and solar power is fluctuating intensely as a function of time and in no way correlated with the energy demand, makes keeping the power grid stable one of the most significant challenges in the future. The lack of correlation is illustrated in Figure 13.2, where total energy demand and the windpower contribution are shown together.

Several options exist to combat the lack of correlation between renewable energy production and energy demand:

(a) Adjusting the power production by on/off control of energy sources, i.e. mechanically halting windmills;

(b) Regulating the power demand to the power production by introducing central energy consumption control mechanisms thereby imposing restrictions on which periods of the day private households can activate extremely energy consuming devices;

(c) Utilizing energy storage facilities thereby eliminating the need for real-time correlation between energy production and demand and

(d) Changing consumer habits.

There is no doubt that the solution is to combine these alternatives in the future energy production infrastructure.

Today, the energy consumption in private households is almost 30% of the total consumption in Denmark [31] and the costs of private energy consumption have almost doubled since 1990. Quite a substantial amount of intelligent electronic control equipment exists in private households today assisting to bring down the energy consumption. Unfortunately most of the equipment focuses on solving a small part of the overall problem and is difficult to setup and use efficiently. Interoperability between devices from different systems is to a great extent non-existing and many different standardized and proprietary communications protocols are used.

Making efficient Information and Communication Technology (ICT) solutions for energy consumption control in private households in the future scenario outlined here is a challenge that requires awareness of the complexity of the overall problem as well as understanding the consumer – being the user of the systems – perspective. The latter includes the necessity to focus on cost.

In Section 13.2 we will elaborate on the challenges related to creating ICT systems which support several technologies in a way that is transparent to the user. Section 13.3 emphasizes the need for a more holistic approach by presenting some of the presently available ICT standards and future technology trends. In Section 13.4 we present our proposed solution to some of the outlined challenges. The actual implementation of the solution, which is in the process of being tested in a private household in Denmark, focuses on solving the interoperability challenge as well as minimizing the efforts necessary to configure the home automation system in a private household. Finally Section 13.5 offers a conclusion on the topics covered in the chapter.

13.2 Challenges

Taking a look at the present level of integration of energy consumption management in private households as an example, it becomes evident that the level of penetration is low.[1] What is the reason for that? The question is simple but the answer is far more complex. However key elements in the answer would be: (a) Low level of consumer cost-benefit awareness, (b) overwhelming number of stand-alone solutions available only providing part of the solution, (c) technical knowledge required to install and operate main part of products, (d) conservatism in the construction sector and (e) high implementation cost.

Addressing these key issues calls for a much more holistic approach when designing energy consumption management systems for private households. This approach should focus on providing an overall solution taking into account the customer and the user needs.

13.2.1 Present Home Automation Approaches

Although politics play an important role resolving (a) and (b) in Section 13.2, a lot can be done from a communication perspective to reduce (or remove) the technically related barriers (c), (d) and (e). Let us have a look at a generic home, where the implementation of energy management is considered. The areas of interest, sensor types and a subset of communication standards used in commercially available solutions today are shown in Figure 13.3.

From a consumer point of view this is far from optimal. A myriad of commercial products are available using yet another myriad of communications standards mixed with proprietary solutions. Very few products aim at integrating control of central heating, airconditioning/climate control and lighting. Scalability is an issue in almost every solution available: if scalability is offered, severe restrictions are imposed on the choice of extra components. Even components based on the same standard can be a no-go due to the fact that deviations from standard profiles are made to a non-neglible extend. Sometimes this can be excused by the fact that no suitable profile has been developed for the application at the time of product development, but most of the times it is justified from an optimization point-of-view (connection time, overall performance, etc.). Choice of wired/wireless communications standard is often driven by the desire to maximize range/coverage and minimize battery power consumption. Thus rendering the different stand-alone

---

[1] Compared to IT equipment like computers, etc. [5].

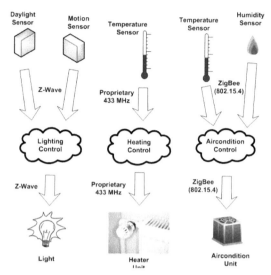

Figure 13.3 Illustration of todays control systems in residential home.

solutions interoperable and thereby not supporting the actual overall goal to make working and transparant energy consumption management system solutions.

Asking the average non-technical consumer to make the right future-proof decision when aquiring energy management systems is not fair. The consequence of asking consumers to navigate in the myriad of solutions is that they refrain from making decisions at all. No-one wants to make investments in equipment that has a high probability of being obsolete in a few years or simply just incompatible with future improvements.

13.2.2 At the Crossroads – Choosing the Future Strategy

Taking the holistic approach in defining the future communications infrastructure for use in connection with energy management systems brings forward a set of "mission critical" elements:

- The consumer perspective – focus on the user.
- Increasing lack of real-time correlation between energy production and consumption.
- The myriads of ICT standards - interoperability.
- Intelligent use of energy production data in the internal private household system.

The focus of present solutions has to a great extend been on the technical challenges, but the user is very important when designing ICT-based equipment for use in private households. Besides technical requirements such as range and battery lifetime, the following requirements are important in order to ensure the development of sustainable solutions:

Easy to Setup and Use: Pairing devices today can be quite tricky. Some devices come out of the box pre-paired with each other and the pairing cannot be changed. This should be made easier in future systems. Daily use of the systems and simple tasks like changing system parameters like room temperatures needs to be straightforward. Understanding of user behaviour and requirements must be included when designing Man-Machine Interfaces and Graphical User Interfaces (MMIs/GUIs) for these systems.

Scalability and interoperability: Adding new sensors and controllers to the system and the initial configuration of such new units should be painless – ideally autodiscovery and autoconfiguration is preferred. Even the addition of new types should be possible. The latter requires a higher degree of system HW/SW flexibility adding a kind of forward compatibility to the systems.

High Level of Quality of Service (QoS): The systems should provide this automatically and not leave the consumers with range problems and a great deal of trial-and-error fault-finding. This could be obtained through adaptive utilization of frequency bands and/or communications standards, i.e. by using Software Defined Radio (SDR) and/or Cognitive Radio (CR) techniques. The consumer should be able to monitor system performance effortless – it is psychologically important to provide this kind of system justification data in order to maintain consumers system confidence.

End-2-End Security: The systems should have no security limitations in the applicability of the transport solution. This is essential not only to maintain good system performance but it is a prerequisite for sustaining consumer confidence.

In addition to these requirements is still the non-negotiable demand for low-cost solutions. This combined set of constraints should lead to implementations of these future communications infrastructures with a high level

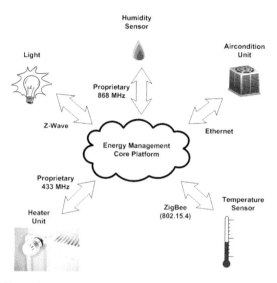

Figure 13.4 Illustration of generic energy management setup in a residential home.

of integration, flexibility and robustness rendering a setup as the one shown in Figure 13.4 possible.

Several "perfect" and "near-perfect" solution alternatives exist, ranging from the idealized setup shown in Figure 13.4 where all units can be handled by the core, to a solution where all controls and sensors strictly adhere to one common defined standard.

The increasing lack of real-time correlation between energy production and consumption needs to be addressed on at least a national basis. Coordination on this level is required both with respect to deciding what data to make available to the private households and when it comes to implementing energy storage facilities. Examples of data that would be highly relevant to private households is the origin of the available energy (renewable energy or fossil fuel) and the price. Aside from supplying this data on a real-time basis the energy suppliers should also provide forecasts thus enabling the private household energy consumption management systems to do intelligent planning. Such planning could include scheduling the start-up of energy-demanding appliances outside the energy consumption peak periods thus providing an averaging effect and making better use of the surplus renewable energy. This would even have a positive impact on the overall energy consumption economy for the private households. The high degree of motivation provided by having a clear impact on economy could further be enhanced by revising energy consumption taxation so it encourages this kind of consumer behaviour.

Interoperability is a growing challenge on the home automation scene. In the next section we will take a quick look at the status of the subject before presenting a detailed description in Section 13.4 of how the challenge is dealt with in our practical implementation: MCHA – Minimum Configuration Home Automation. This implementation addresses several of the key elements outlined in the present section as crucial for designing sustainable solutions.

13.3 ICT Standardization and Technology Trends

Traditionally the house of tomorrow is considered the most obvious example of where wireless multimedia meet wireless sensor networks. The connectivity to and from the house is traditionally connected via a gateway to the local house-LAN sensor based network inside the house.

Despite this traditional gateway-based view of the wireless network infrastructure there are, as stated above, other initiatives attempting to make a more future proof and technically "clean" solution based on the Internet technology (see Figure 13.5). If it could be possible to embed the sensor network of the house directly to the outside Internet protocol based connectivity network end-2-end security can be achieved. Additionally there will be no problem with scalability of number of sensors or amount of traffic. In principle the classic hour-glass model of the Internet protocols allows for any type of underlying link layer technology. It is independent of the physical link layer technology, which could be Bluetooth, Z-Wave, Zigbee or any other similar standard. Most importantly, for home automation, the plug-and-play can easily be implemented using well proven Internet application technologies. A supporting factor regarding how well minimum/zero configuration can be achieved based on such a pure IP-based infrastructure is that a number of higher layer (OSI reference model) initiatives supports the concept perfectly.

More than 800 leading companies in computing, printing and networking; consumer electronics; home appliances, automation, control and security; and mobile products [23] have created two related standards: Universal Plug And Play (UPnP) and Digital Living Network Alliance (DLNA) [4] which is based on the UPnP standard. The goal is to allow devices to connect seamlessly and to simplify network implementation in the home and corporate environments and to offer simple and seamless access to digital content. In order

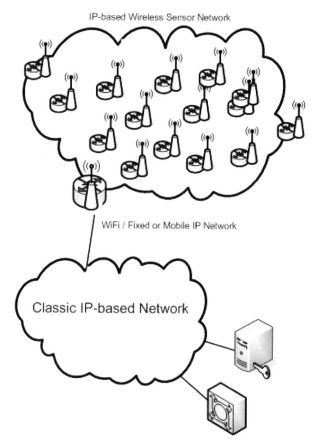

Figure 13.5 IP-based wireless sensor network connected to an IP infrastructure.

to guarantee interoperability, products must be designed in respect to some device guidelines specified in the standards. The architecture described in the standards support zero configuration networking and automatic discovery. This means that a device dynamically can join a network, get an IP address, announce its name and capabilities upon request, and learn about other nearby devices. The main features are based on technologies such as IPv4, IPv6, XML and SOAP to achieve this. DLNA and uPnP are mainly suitable for multi-media devices, since the use of XML and SOAP puts to much strain on the computing power of small sensor devices.

13.3.1 Minimum Configuration – On the Network Layer

Until recently the Internet Protocols (IP) have been considered too large, complex and power consuming (processing, memory, etc.) to be put into wireless sensor networks efficiently. Typically there are very strong requirements on low power and low complexity in these wireless sensors.

Recently it has been shown how it is possible to implement an IP-based solution given these requirements. Currently the Internet Engineering Task Force (IEFT) is standardizing a solution attempting to extent the Internet architecture to wireless sensor networks. For wireless sensor networks the IEEE 802.15.4 standard [11, 22] is currently found favorable for the physical and link layers. The IP solution on top is based on the 6LoWPAN[2] currently addressed in the RFCs 4919 and 4944 [16, 18].

The idea of 6LoWPAN is to offer an IPv6-based solution for the critical embedded wireless requirement of high reliability and adaptability, long lifetime on limited energy and within highly constrained processing resources to minimize cost. The 6LoWPAN based sensor network should use below 1% of the 802.11 power (and additionally be turned off most of the time) and be running on a small 8 bit microcontrollers with only KBs of memory.

Additionally the network should be manageable even with many devices. IP is not just used in a specific application context like the intelligent house, but also for the general case. That is, provide interoperability with all other potential IP network links, offer the potential to route to any IP-enabled device within an offered security domain, co-exist with other systems and yet provide a very robust operation. In that sense the IP-based sensor network using 6LowPAN is much more general than a feature to be used for sensors in your home, it can actually be seen as one of the drivers for the vision of *The-Internet-of-Things* or the *The-Internet-of-Everyday-Things* [21] where physical devices of all sorts are connected using the Internet protocols over wireless as well as wired links.

The IP-based sensor network protocol solution is illustrated in Figure 13.6 with the 6LoWPAN protocol between the link and network layers of the sensor network. The figure shows the principle of the protocol layering when connecting a wireless sensor network using 6LoWPAN and 802.15.4 and some more fixed access to the Internet through either WiFi, mobile broadband or similar. No gateway is needed to provide the Internet connectivity. It is clear that the concept is an extension to the IP technology suite that provides a new class of devices, but in a familiar (to traditional IP) way.

---

[2] IPv6 over Low-Power Wireless Personal Area Networks.

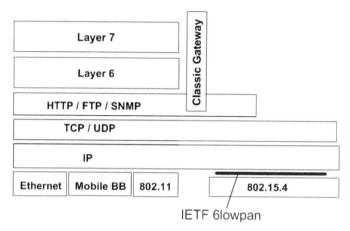

Figure 13.6 Protocol stacks with IP for both the access network and the wireless sensor network based on 6LoWPAN [6]. No protocol translation gateway is in principle necessary.

Regarding 6LoWPAN IETF address a number of issues that needs to be solved. For standardization the focus will be to ensure interoperable implementations of 6LoWPAN networks and to define the necessary security and management protocols and constructs for building 6LoWPAN networks, paying particular attention to protocols already available. Details on the current IETF work can be found in [12].

There are a number of evident indications of both commercial products as well as open source components implementing an IP-based solution for wireless low power sensor networks. One open source example is Contiki [3], a portable, multitasking operating system for memory efficient networked embedded systems and wireless sensor networks. It is designed for configurations of 2 KB of RAM and 40 KB of ROM. An example of a company offering a commercial product using 6LoWPAN is ArchRock, which have at least two products available. An IP-enabled wireless sensor network system that can be accessed through a web interface and additionally an energy optimizing package that enables organizations to zoom in on the biggest energy consumers in e.g. buildings [26].

Additionally a number of standardization organizations are considering adopting 6LoWPAN over IEEE 802.15.4 for the wireless sensor networking. One example is the North American International Society of Industrial Automation (ISA), a nonprofit organization standardizing automation [20]. The standard, ISA 100.11a [29], is a wireless networking technology designed to be open and available via the Internet. The standard is still in the early stages and based on IEEE 802.15.4 with frequency hopping and mesh routing.

Given the amount of work in progress and the attention on sensor network applications there is no doubt that low-power IP-based solutions will find its way onto the home automation scene.

13.3.2 Minimum Configuration – On Radio Link Layer

Equivalently to the IP-based connectivity described above, a number of initiatives also supporting the minimum configuration or zero configuration for home automation is also proposed for the physical radio link layer.

The concept of software defined radio (SDR), defined as "A radio in which some or all of the physical layer functions are software defined" support the fundamental minimum configuration concept. In principle SDR defines a collection of hardware and software technologies where some or all of the radio's operating functions (also referred to as physical layer processing) are implemented using FPGAs (Field Programmable Gate Arrays), DSPs (Digital Signal Processors), GPPs (General Purpose Processors) or SoC (System on Chip) specific programmable processors [24]. In other words each transceiver can, in real-time, select between a number of available pre-defined and implemented radio front-ends (see Figure 13.7), such as 802.11, 802.15, a proprietary ISM band radio link or something similar.

There has been a significant amount of research on SDR, but today SDR is still in the early stages of commercial usage. Regarding standardization initiatives have started by ETSI designing an architecture for reconfigurable radio systems [2]. The interest in SDR accelerates when combined with other types of intelligent radio concepts such as Adaptive Radio (AR), Cognitive Radio (CR) and Intelligent Radio (IR) [10,25,30]. In a sense SDR can be seen as an enabling technology for these next generation radios. The generalized block-structure of a SDR/CR type receiver is shown in Figure 13.7.

Regarding these intelligent radios a number of completely unsolved issues has to be dealt with. One example among many is the research area of how to do the sensing and selection of the spectrum. Additionally it seems like the complexity of these systems must be relatively high, complicating the implementation of a very low cost, low power, low processing sensor type of device. This is a theme for further research.

Finally, another parameter emphasizing the importance the intelligent radios is when looking at the possibility of creating a system based (exclusively) on unlicensed spectrum. The classic ISM bands around 2.45 and

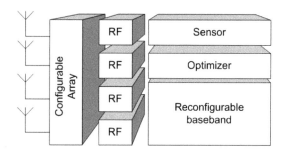

Figure 13.7 Principle block structure of a SDR/CR combined transceiver.

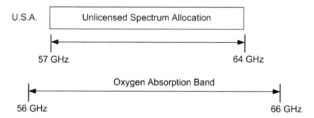

Figure 13.8 Spectrum allocation and the oxygen absorption band.

5.8 GHz [1] are to be used, but to provide significant bandwidth, up to 6 GHz (e.g. to transmit wireless HDMI signals), new spectrum is considered for intelligent houses. Specifically the 7 GHz bandwidth allocated on the 60.5 GHz carrier frequency for unlicensed usage (in USA), see Figure 13.8, opens up for increased possibility and minimizes the spectrum bottleneck. The spectrum around 60 GHz become especially interesting because the oxygen molecule (O_2) absorbs electromagnetic energy at 60 GHz like a piece of food in a microwave oven. This "natural" range reduction can be used for much easier reuse of frequencies since the interference can be controlled more easily.

It should be said that the 60 GHz wireless communication is still on the very early stage, research-wise as well as commercially, the latter being non-existing. With strict regulation on the transmit power level challenges on how to deal with a significant path loss, channel dispersion, robust modulation, etc., are parameters to overcome. Prototype have recently shown possible bandwidths of up to 300 Mbit/s in the 60 GHz carrier frequency range.

Expanding the usable frequency range and incorporation of more intelligence on the radio link layer provide a set of important and extremely useful tools. They will assist in dealing with the increasing challenges with respect to interference and the inherent security issues in wireless communications. This assistance is important in order to be able to provide a high level of QoS in future wireless home automation system.

13.4 One Part of the Solution: MC-HA

Some of the technologies presented in Section 13.3 focusses on interoperability, whereas others focuses on QoS parameters. All of the challenges listed in Section 13.2, has not been considered though, an hence a true holistic solution is still needed. This section presents a practical solution solving not only the interoperability issue, but also many of the other challenges mentioned. The Danish research project called "Minimum Configuration – Home Automation", abbreviated MC-HA. Initially an overview of the project is given, continuing with a description of the work done through user-driven innovation. This is followed by a series of subsections of more technical nature about a home automation framework that has been developed in the project and its characteristics and deployment.

13.4.1 Project Overview

In 2008 the Danish Enterprise and Construction Authority decided to fund the research project MC-HA [17] with a budget slightly below 1 M Euro. The objective of this project is to develop, through user-driven innovation, a unifying concept of how different electronic solutions can be configured in the home, in such a way that they will become applicable and relevant for users. The partners in the project are:

- Aarhus School of Engineering,
- The Alexandra Institute,
- Seluxit, and
- Develco Products.

The project is based on participatory design of a multidisciplinary cooperation in the user innovation [27]. Participatory design is a new approach for a number of the involved industrial partners and they will in the process, both learn to use the new methods, while describing the applicability of them for industry value. The project will generally work with the following two kinds of users:

1. A reference user group consisting of families with or without children.

2. Two families who will live in conventional houses in Denmark and Portugal.

13.4.2 User Driven Innovation

The core of the project is centered around the concept of user-driven innovation [7], where the work has been split into four parallel tracks of work:

Minimum configuration: How can users find out about configuring the optimum behavior of the automation in their own home?

Use energy when cheaper/greener: How can the usage of energy be moved from periods of the day where the demand is highest to periods where there is too much supply of energy?

Visualize energy usage: How can the energy usage in private homes be visualized such that the behavior of the energy usage can be adjusted as a consequence of that?

Light and indoor climate: How can the light and indoor climate of the home be adjusted such that the automation will be working in an optimal fashion for the inhabitants of the home?

An anthropologist is used to drive the user driven innovation tracks here. Initially, this included semi-structured interviews, energy tours in the home, scenario-observations and cultural probes during a period of approximately three months involving 24 families. The initial interviews gave an indication of what might be interesting to further elaborate through observations and cultural probes. Interviews were either recorded on tape or on video, while energy tours and observations were all videotaped. The cultural probe consisted of a ground plan of the users' home, a camera and a diary. The task of the users was to draw their movements in the house on the floor plan, take photos of what their home look like in the morning and in the evening and also describe their activities in the diary.

The use of different methods resulted in collected data which was materialized in many shapes and forms (audio, visual, textual and material) each giving a different feel for, and various perspectives of, the complex life of the participating users. The main challenge is how to make most effective use of the data that has been collected in the design of solutions for real users.

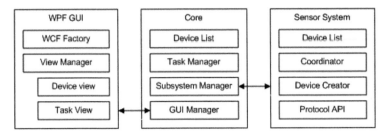

Figure 13.9 The EPIC framework system overview.

13.4.3 The EPIC Framework

Based on the user driven innovation, a home automation framework has been created, called "Extendible Protocol Independent unit Controller" (EPIC). Once matured, this framework will be open source for anyone to use. The first version of the framework has the following focus points:

Protocol Interoperability: The framework must be able to handle several physical layer communication protocols in a way that is transparent to the user.

Solid Backbone: Since it is the idea that several third party developers will make additions to the framework in the future, it is very important that the framework core is very stable. This will ensure a robust system.

Test Platform: The user driven innovation will create a lot of feedback on how the home automation system should work. It is essential that the framework enables rapid prototype testing of several user interfaces, to support the frequent user feedback.

13.4.4 The EPIC Framework Structure

The EPIC framework developed as a part of the MC-HA project has high focus on reliability and the ability to adapt different protocols into a single system which is easy to configure for the user. An overview presenting the main parts of the system is shown in Figure 13.9.

The system is comprised of the following main components:

The Core is the main software module which controls devices and holds abstract representations between devices in the system (explained more thoroughly in Section 13.4.5). Here the relations between devices are represented as Tasks and evaluated upon state change of a physical device. The core has two interfaces which are exposed; a Subsystem Manager and a GUI Manager. The Subsystem Manager interfaces to the Sensor System which is responsible for a single protocol. The GUI Manager interfaces the Graphical User Interface (GUI). Both interfaces are exposed through Windows Communication Foundation (WCF) [14] which easily provides end-points of types such as .Net remoting, ASP.NET Web services (ASMX), etc.

The Sensor System is a small sub-component which connects to the core. It is responsible for managing a specific protocol such as ZigBee [15] or Z-Wave [9] through the Coordinator. Another important responsibility of the Sensor System, is the transformation between a specific device representation on the protocol level and the abstract device represent-ation used in the Core. Each device is described as a container with a collection of features. Each feature has a detailed description specifying what the given feature provides and how to interact with it. The im-plementation made for ZigBee and Z-Wave is implemented with use of System.AddIn [32] for loading the interface description of the features which a specific device provides. This is done through the Device Cre-ator. This makes the extension of new features easy and introduces a higher level of security and reliability.

The GUI component is constructed as a Windows Presentation Foundation (WPF) [28] application connected to the core through the WCF Factory. To be easily expendable and more robust System.AddIn is used to con-struct a stable plug-in platform. The System.AddIn technology is able to handle a wide range of problems commonly encountered when building a GUI. This means that new views of the system easily can be created and both the security and reliability of the application can be preserved even when third party vendors provide new views. The base function-ality includes presentation of devices in a house along with the device information. The highly important Task management, where different devices of the system can be related to each other at the abstract level, is also presented here. To aid the pairing of devices into Tasks a range of predefined device relations is used in order to automate the process.

13.4.5 EPIC Protocol Interoperability

One of the main objectives of the EPIC framework is to ensure interoperability between several communication protocols in a way which is transparent to the user. This will provide the user with a broader choice when it comes to purchasing home automation devices, and hence make home automation more accessible to a wider part of the population. This, in turn, will help to lower the general burden on the environment, by reducing the CO_2 emission.

In order to support several coexisting protocols in the EPIC framework, the different subsystems and the system core is decoupled. The system core is operating on an abstract representation of home automation devices with a range of services describing the unit and how to interact with it. This abstract representation is based on the highest common denominator of several analyzed communication protocols. When the system needs to send a command to a device, an abstract message is produced based on the features supported by the device. This message is sent via the Subsystem Manager to the coordinator of the concrete subsystem, where the message is translated into the protocol specific equivalent of the abstract message. Using the protocol API, the newly translated message is sent to the hardware interface and out on the wireless network. When a message is received on the hardware, the opposite translation is carried out, in order to create a message the system core can understand and operate on. This transformation must be done near the protocol since the transformation in both directions needs detailed protocol information.

Figure 13.10 shows an example of the transparent protocol interoperability. A ZigBee motion detector device reports that is has sensed movement. The message is received by the physical ZigBee coordinator since it is a generic ZigBee command, and is passed to the ZigBee Sensor System. Here, the message is translated to an abstract representation and passed on to the EPIC Core. The Core evaluates all the tasks configured into the system, and finds a relation between the state change of the abstract device and a feature of another abstract device. A new abstract message is created by the Core, and sent to the Subsystem Manager, which identifies the communication protocol of the destination device to be Z-Wave. Hence, the Z-Wave Sensor System receives the abstract message, and translates it to a generic Z-Wave message. Finally, the message is sent via the physical Z-Wave coordinator to the destination end device.

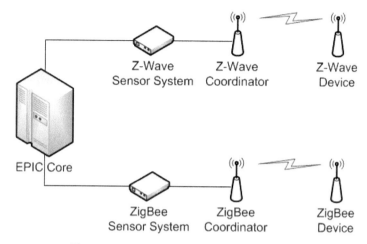

Figure 13.10 Protocol interoperability example.

13.4.6 Reliability

Reliability is an important issue in any system, especially for a home automation system. While the failure of such system might typically not cause property loss, it might lead to severe loss in confidence by the users. Without confidence in the system that control their homes, the users will most definitely not be willing to use it.

The system was from the beginning designed to be open and easy expandable by allowing plug-ins made by anyone. These plug-ins could be either new GUI views or protocols and device types. This starting point implies that in the future, the system will grow and eventually deploying buggy plug-ins in the system. Some effort has therefore been put into ensuring that these buggy plug-ins will not crash the entire system leaving it useless.

The first step to achieve the desired level of reliability was to separate and distribute the system into a series of smaller subsystems that would be able to minimize the loss in case one of the subsystems fails. For this matter, it was chosen that a separation must exist between the Core and the different Sensor Systems. The system cannot ensure that a Sensor System will not fail, but in that case, the Core will still be able to communicate to the GUI that one of the Sensor Systems has failed.

The system is not only open to the creation of new Sensor Systems by third party developers. It is also possible to extend the current Sensor Systems as it might be useful in the case of, for example, a new type of device being created for a protocol that its Sensor System does not support. These extensions, could also potentially contain bugs crashing a Sensor System. Therefore System.AddIn is used to add these new plug-ins to the system running in separate processes. In case of the detection of a faulty plug-in, it can simply be unplugged because System.AddIn uses the kernel to isolate these software components from the rest of the system. As a consequence, the Sensor System will still be up and running in the case of a fault.

The GUI is also made to be expandable and because of the user driven innovation, it is highly desirable that it is easy to do so. With System.AddIn, it is possible for plug-ins to provide their own representation on the GUI and the defined interfaces ease GUI expansions for third party developers. GUIs are typically unstable and again, it is not desirable that the whole GUI shuts down because one of the plug-ins has crashed. So also in this case System.AddIn can help providing the reliability that the system needs.

13.4.7 Framework Security

A very important aspect for realistic deployment is to have security properly implemented. It is necessary to ensure that intruders cannot access the EPIC systems opening up the control of the household. Note that security in relation to private homes is not just a matter of adding encryption to the different protocols used. The EPIC framework incorporates several protocols with different levels of security - all messages in the ZigBee protocol are encrypted, whereas only a selected subset of the Z-Wave protocol used to control devices such as door locks is encrypted. If the Z-Wave network should be compromised, is it important to ensure that the intruder cannot gain control of any of the secure ZigBee devices through the use of an unsecure Z-Wave device. These types of security measures are not yet incorporated in the EPIC framework in a fully satisfactory fashion.

In addition to potential security breaches in the wireless protocols used, it is also important to ensure secure network communication between the different distributed parts of the EPIC framework. This threat could be solved by using standard components like Secure Socket Layer (SSL) between the EPIC core, GUI and subsystems. However, to appropriately support security it is necessary to analyze all possible threats and subsequently come up with mitigation plans against all these threads.

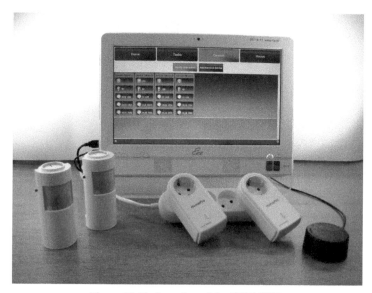

Figure 13.11 Snapshot of the EPIC GUI.

13.4.8 The Existing EPIC Prototype

The EPIC framework will become an open source project, since it is the plan that third party developers will aid in the expansion of the framework, in order to support more underlying physical communication protocols in the future. As a result of this strategy, it is very important that the system is completely reliable, and absolutely essential that no "single point of failure" exists, as stated earlier. This has been achieved by separating the system core from the GUI and Subsystems. This ensures that even though a protocol has minor errors, the rest of the system can still run without any problems. In addition, System.AddIn is used in order to dynamically load the different GUI views and device features in their own application domain. By separating these unreliable parts from the core of the system, a very reliable and robust system has been created.

The EPIC framework will, as part of the MC-HA project, be installed in houses in Portugal and Denmark. After a thorough testing phase, and once the framework is deemed stable it will be made available at SourceForge.

The initial results of this deployment have detected a number of challenges that will need to be carefully considered in order to enable average non-technical users to configure their own homes. These challenges includes:

- Devices that use wireless technologies have different ways of identifying themselves to a controller. Holding a button pressed for a specific amount of seconds to make a pairing of devices is non-trivial for non-technical stakeholders.
- Wireless controllers have limitations that are hard to efficiently communicate to non-technical stakeholders. For example a ZigBee coordinator is able to deal with 14 end devices and 6 additional routers. This means that one cannot connect 15 devices without a special way of setting up the system.
- The devices for home automation are not yet produced in large numbers and as a consequence the reliability of many of the devices can be relatively low.

In addition there is no doubt that in general it is necessary that the prices for sensors and actuators to be used in private homes needs to be lowered significantly in order to make it economically viable for private people to invest in this kind of technology.

13.5 Concluding Remarks

In this chapter we have described how the energy consumption of private households constitute a significant amount of the overall power consumption budget. For the last 5–10 years a myriad of different intelligent ICT technologies has evolved to control and regulate the energy consumtion of a normal household. Despite that home automation has not gained a market share that has any significance. It is too complicated and complex to combine and configure the systems. We have presented a number of proposals on some underlying technologies that might help out.

Additionally it is important to realize how the heavy expansion with respect to renewable energy (e.g. in the form of wind turbines) re-defines the "game" and calls for re-thinking the CO_2 strategies. First of all, home automation needs to provide solutions for local control as well as intelligently communicating with a distributed power grid. Thinking home automation and energy grid management as a combined challenge is a necessity. We furthermore believe home automation applications should be the result of multidisciplinary work: Civil engineers working together with ICT engineers, which could add building construction and material knowledge to the control implementation process resulting in solutions that actually do the job.

The MC-HA project and EPIC framework is an attempt to solving the problem of how to take todays existing products, using various types of wireless techniques, usable in a kind of plug-and-play manner. In other words a proposal on how to ensure intelligent home automation takes off. Future work involves the introduction of external data containing information of the source of energy and price, and based on these informations, as well as the users preferences, the framework will suggest better energy usage. An example of such a suggestion would be to postpone doing the laundry until a surplus of renewable energy is available.

Finally, it should be emphasized that it will be important to respect the consumer preferences and limitations, why these parameters should play an important role in the development and roll-out of future home automation ICT solutions.

References

[1] The Industrial Scientific and Medical (ISM) radio bands, `http://en.wikipedia.org/wiki/ISM\_band`.

[2] ETSI TR 102 680 V1.1.1 (2009-03). Reconfigurable Radio Systems (RRS); SDR Reference Architecture for Mobile Device, March 2009.

[3] Dunkels, A., Grönvall, B. and Voigt, T., Contiki – A lightweight and flexible operating system for tiny networked sensors. In *Proceedings of the First IEEE Workshop on Embedded Networked Sensors*, Tampa, Floria, USA, IEEE, November 2004.

[4] Digital Living Network Alliance, `http://www.dlna.org`.

[5] Park Associates, Media Servers in the Digital Home. White Paper, 2006.

[6] Culler, D., Secure, low-power, IP-based connectivity with IEEE 802.15.4 wireless networks. Industrial Embedded Systems, 2007.

[7] Entwistle, J.M. and Astrid Søndergaard, A., Changing the role of the anthropologist in user driven innovation processes. From data collector to process facilitator. In *Ethnographic Praxis in Industry Conference 2009*, Chicago, August 2009.

[8] Nordel Organisation for the Nordic System Operators, Energy balances 2012, 2009.

[9] Galeev, M., Catching the Z-Wave. *Electronic Engineering Times India*, pp. 1–5, October 2006.

[10] Haykin, S., Cognitive radio: Brain-empowered wireless communications. *Selected Areas of Communications, IEEE Journal* 23(2), 201–220, 2005.

[11] IEEE, IEEE Standard for information technology – Telecommunications and information exchange between systems – Local and metropolitan area networks – Specific requirements Part 15.4: Wireless Medium Access Control (MAC) and Physical Layer (PHY) specifications for low-rate Wireless Personal Area Networks (WPANs), IEEE Std 802.15.4-2006 (Revision of IEEE Std 802.15.4-2003), September 2006.

[12] Internet Engineering Task Force (IETF), IPv6 over low power WPAN (6LoWPAN) charter, `http://www.ietf.org/html.charters/6lowpan-charter.html`, 2009.

[13] IPCC, Climate change 2007: Synthesis report. Technical Report, IPCC (Intergovernmental Panel of Climate Change), Geneva, 2007.

[14] Löwy, J., *Programming WCF Services*. O'Reilly, February 2007.

[15] Kinney, P., ZigBee technology: Wireless control that simply works. In *Communications Design Conference*, pp. 1–21, October 2003.

[16] Kushalnagar, N., Montenegro, G. and Schumacher, C., RFC-4919, IPv6 over Low-Power Wireless Personal Area Networks (6LoWPANs): Overview, Assumptions, Problem Statement and Goals. IETF Request for Comment, 2007.

[17] Minimum Configuration – Home Automation (MC-HA), http://www.iha.dk/minimumconfiguration, 2009.

[18] Montenegro, G., Kushalnagar, N., Hui, J. and Culler, D., RFC-4944, Transmission of IPv6 packets over IEEE 802.15.4 networks. IETF Request for Comment, 2007.

[19] OECD/IEA, World energy outlook 2008: Executive summary. Technical Report, International Energy Agency, Paris, France, 2008.

[20] The International Society of Automation, http://www.isa.org/.

[21] Internet of Things/Internet of Everyday Things, http://en.wikipedia.org/wiki/InternetofThings/.

[22] ZigBee Standards Organization, Zigbee specification, version r17, January 2008.

[23] Universal Plug and Play Forum, Upnp device architecture 1.1, 2008.

[24] SDR Forum, Promoting the success of next generation radio technologies, http://www.sdrforum.org.

[25] Hoffmeyer, J.A., Venkatesha Prasad, R., Przemyslaw Pawelczak, P. and Berger, H.S., Cognitive functionality in next generation wireless networks: Standardization efforts. *IEEE Communications Magazine*, April 2008.

[26] Arch Rock, http://www.archrock.com.

[27] Schuler, D. and Namioka, A. (Eds.), *Participatory Design: Principles and Practices*. Lawrence Erlbaum Associates, 1993.

[28] Sells, C. and Griffiths, I., *Programming WPF*. O'Reilly, September 2005.

[29] ISA Standardization, ISA100.11a, Wireless systems for industrial automation: Process control and related applications, 2009.

[30] Staple, G. and Werbach, K., The end of spectrum scarcity. Technical Report, IEEE spectrum, 2004.

[31] Danish Energy Agency Statistics and Indicators, Energy statistics 2007. Technical Report, Danish Energy Agency, 2008.

[32] CLR Inside Out: .NET Application Extensibility, http://msdn.microsoft.com/en-us/magazine/cc163476.aspx, 2007. MSDN – Developers Center – MSDN Magazine – Home – Issues – 2007 – February – CLR Inside Out: .NET Application Extensibility.

14

Photovoltaic-Aware Multi-Processor Scheduling – A Recipe for Research

Peter Koch

Department of Electronic Systems, Aalborg University, 9220 Aalborg, Denmark

Abstract

In recent years, the number of programmable processing elements which can be integrated into a self-contained battery powered embedded system has increased substantially. Not only can multiple processors alleviate the growing need for higher computing capacities, it can also provide a convenient mean for a potential reduction in the overall energy consumption in such systems. Considering the lack of improvement in battery technologies, this chapter discusses the foundation for taking advantages of energy harvesting in a multiprocessor based embedded system.

Keywords: Energy harvesting, multiprocessor systems, scheduling, resource optimization, battery powered devices.

14.1 Introduction

In high-performance embedded wireless systems such as mobile handsets, PDAs, handheld gaming consoles, portable medical devices, and sensor motes, the battery lifetime is an increasingly important design parameter due to the fact that long and reliable up-time provides comfortable use, and in many cases also leads to reduced maintenance expenses for the end-user. To a certain extent the growing requirement for prolonged battery lifetime is indirectly accomplished by the evolution in chip and packaging technology which

R. Prasad et al. (Eds.), Towards Green ICT, 211–228.

enables smaller and smaller transistor geometries paving the way for more energy efficient computing platforms. Although predicted to remain valid for at least yet another decade, Moores law however, does no longer compare favourably to the constantly increase in the computational complexity we are witnessing for advanced embedded applications. At the same time, state-of-the-art battery technologies are also known to lack sufficient improvements in order to counteract the growing energy consumption needed by the growing algorithmic complexity.

One interesting approach which may help alleviate the declining effect of Moore's law is to equip the target architecture with multiple processing engines (PEs), thus enabling the possibility to trade off execution speed and consequently power consumption for more hardware units. According to the most recent International Technology Roadmap for Semiconductors, The ITRS 2007 Edition [1], the number of PEs which typically will be found on application specific Systems on Chip (SoC) is expected to show an almost polynomial grow rate in the coming decade, reaching nearly 1500 PEs in year 2022. Similarly, the upcoming field of GPU computing (Graphical Processing Unit), provides several hundreds of individually programmable cores on a single chip. Examples are the state-of-the-art NVIDIA Tesla GPU, which has 240 cores, and the NVIDIA's next generation GPU architecture, known as the Fermi GPU, which implements approximately three billion transistors, featuring up to 512 cores [2].

For time-constrained algorithms having a certain amount of inherent parallelism, their execution on multiple programmable processors may provide an overall energy saving due to the quadratic nature of the power/supply-voltage relation characterizing CMOS technology (Complementary Metal Oxide Semiconductor) which is still considered the semiconductor "working-horse". By distributing the computational load onto multiple cooperating processors, the supply voltage (as well as the clock frequency) can be reduced, still however, maintaining the overall computational capacity due to the increased number of processing units. The power saving caused by the V_{dd} reduction will cancel out many-fold the linear power consumption increase encountered by up-scaling the number of processors, thus providing a net power saving.

At the same time, the system designer should also be aware that state-of-the-art rechargeable electrochemical batteries, typically used in embedded wireless devices, have non-linear behaviour which means that the overall available energy capacity may vary according to different discharging profiles. Basically, a dynamic discharging scheme, i.e., interrupted or load-varied which enables the battery to recover, may help extending the battery lifetime. Numerous researchers have addressed the problem of conducting energy-aware scheduling in multiprocessor environments using appropriate voltage- and frequency scaling techniques, but few, however, have considered the challenge to take simultaneous advantage of using multiprocessor topologies and e.g., advanced discharging techniques in order to prolong the battery lifetime.

With the prospect of still more complex and thus energy hungry portable wireless devices, the user on the move, who typically has limited or troublesome access to wall sockets, may want to take advantage of the possibility to partly recharge the batteries from environmentally harvested energy. An interesting example of this concept is solar cell powered handsets which were commercially introduced to the market in 2009 by Japanese Sharp, and Korean Samsung, and LG. Natural or artificial light is an almost everywhere accessible energy source, and therefore it also makes much sense that the first Energy Harvesting (EH) based handsets to hit the market are based on photovoltaic (PV) havesting techniques. It is notable however, that for such handsets the typical yield is approximately a 1–2 minutes talk time after being exposed by full sun for several dozen minutes.

In conclusion it is therefore suggestive to observe that the design of future energy-aware wireless embedded systems needs to address a variety of different disciplines which cannot be managed independently, but rather should be combined in order for the designer to achieve at optimal or near-optimal solutions. Consequently, with the overall purpose of visualizing where we are possibly heading in terms of design methods for low energy embedded wireless systems, and how such methods may potentially interact, this chapter discusses fundamental issues in multiprocessor scheduling, photovoltaic EH, rechargeable batteries, and in particular what are the research possibilities and challenges when combining such technologies in energy optimal embedded wireless system design.

14.2 Multiprocessor Scheduling Overview

Architectures for modern embedded wireless systems are rapidly shifting towards concepts based on e.g., Software Defined Radios (SDR) where the majority of the radio functionality such as down-conversion, filtering, channelizing, and baseband processing, is conducted digitally, i.e., in terms of time- and amplitude discrete signals. Interesting features of such digital signal processing (DSP) algorithms are (1) that they normally have no or very little inherent decision making which means that they have a (quasi-) static nature enabling compile-time scheduling, and (2) that in many cases they exhibit a high degree of inherent parallelism. Implementation of such algorithms on a multiprocessor architecture is a three-fold combinatorial optimization problem, including (i) assignment of the arithmetic operations to the individual processors, (ii) sequence ordering of the operations on the processors such that given precedence relations are maintained, and (iii) calculation of the actual start time of the operations taking into account the status of the Inter-Processor Communication (IPC) hardware. This is done subject to minimizing a given objective function, normally the total execution time which is also known as the *makespan*.

Due to no or little inherent decision making, DSP algorithms are mostly *data independent* computations which execute iteratively once every sample period T. For each such period the algorithm can be described in terms of a data-flow based graph notation, e.g., the *Synchronous Data Flow* (SDF) paradigm [3]. This notation serves two purposes. First of all it captures the number of data tokens being produced and consumed by the individual operations, and secondly it can be used to calculate conveniently the number of times each operation has to be invoked within every iteration. This enables a reliable conversion of the SDF graph into a more suitable graph notation, normally denoted *Directed Acyclic Precedence Graph*, DAPG, which is next used as input to the heuristic that performs the static scheduling [4]. In the following, we give a formal definition of the scheduling procedure.

The DAPG is denoted $G = (S_n, S_a)$ where $S_n = \{n_i : i = 1 \dots N\}$ is a finite set of *nodes* representing the computations (the grains) in the algorithm, and $S_a = \{(n_i, n_j) \mid n_i \rightarrow n_j\}$ is a finite set of *arcs*, indicating the directed *precedence links* among the nodes. Associated to every node n_i is a weight w_i denoting its *estimated run time* (in a homogeneous multiprocessor environment), and also to each precedence link is assigned a weight factor c_{ij} which specify the amount of *data tokens* transferred from n_i to n_j. This model can represent nodes of possibly different grain sizes.

Concerning the architecture which is a collection of interconnected processors, it can be defined as the set $S_p = \{p_k : k = 1 \dots P\}$, where P typically is less than 6–8 for the type of embedded systems we are addressing in this work. We assume that the processors can be organized in a *fully interconnected topology*, i.e., a zero-hop configuration where all processors can communicate directly with any other processor. For the relatively small maximum number of processors, this is a realistic assumption, for message passing as well as for shared memory configurations. We also assume

(i) that the "one-token" IPC time is *identical* and *constant* between any two processors, and (ii) that the *Intra-Processor Communication* time equals zero.

Multiprocessor scheduling can be defined as the set $\mathcal{F}$ of functions g which are able to map G onto an architecture S_p. Now, $g(n_i)$ returns the *set of processors* p_k on which n_i is scheduled. This notation is applied since we, in general, allow n_i to be *recalculated* on multiple processors in order to possibly minimize the amount of IPC needed. Next, $\forall g \in \mathcal{F} \; \exists \; \bar{g}$, where $\bar{g}(p_k)$ is the set of nodes mapped onto processor p_k by the function g. Similarly, $\forall g \in \mathcal{F} \; \exists \; S_g$, where S_g denotes the set of valid schedules, s, which can be obtained using g. Therefore, $S_g \subseteq S_{total}$, where S_{total} is the set of all possible valid mappings of G onto S_p, i.e., the *solution space*. Given a schedule $s \in S_g$, if $n_i \notin \bar{g}(p_k)$ then $s(n_i, p_k)$ is undefined, otherwise it denotes the invocation time of n_i on processor p_k.

Employing these definitions, a schedule s generated by g is *valid* if the following conditions are true for all pairs of nodes $(n_i, n_j) \in S_a$.

$$\forall p_l \in g(n_j)$$
$$\text{if } p_l \in g(n_i) \text{ then } s(n_i, p_l) + w_i \leq s(n_j, p_l) \tag{14.1}$$
$$\text{if } p_l \notin g(n_i) \text{ then } \exists \; p_k \in g(n_i) \;|$$
$$s(n_i, p_k) + w_i + t_{IPC}(n_i, n_j) \leq s(n_j, p_l) \tag{14.2}$$
$$\forall p_k \in S_p, \quad \forall n_i \in S_n, \quad n_j \in \bar{g}(p_k)$$
$$\text{if } n_i \neq n_j \wedge s(n_j, p_k) \geq s(n_i, p_k) \text{ then}$$
$$s(n_j, p_k) \geq s(n_i, p_k) + w_i \tag{14.3}$$

where (i) we assume that $k \neq l$ and (ii) $t_{IPC}(n_i, n_j)$ denotes the total communication time (including possible time for contention) between the nodes n_i and n_j. Due to the hard real-time constraints normally characterizing DSP algorithms employed in wireless baseband processing, we realistically assume non-preemptive execution of the nodes.

14.3 The PV-Array Circuit Element

For cellular handsets and unattended embedded communication systems such as wireless sensor motes, important design considerations include cost, size, reliability, up-time, as well as maintenance overhead. In terms of the energy supply needed, such design metrics may benefit from PV-EH, but in order to potentially take advantage of this specific energy harvesting concept in a static multiprocessor scheduling environment, our first task is to briefly survey characteristics of PV-array technologies to gain insight into their behaviour.

A solar panel consist of one or more PV-arrays which themselves are composed of PV-cells. It is beyond the scope of this chapter to elaborate extensively on the semiconductor physics which enables the PV-cell to convert irradiated photons to an electric current (and heat), and thus we will limit our discussion to a relatively simple but reliable circuitry model and its corresponding (I, V) characteristic. Figure 14.1a illustrates the circuit equivalent of an ideal PV-cell consisting of a constant current generator I_{PV} and a diode with current I_D. At the output terminals, this circuitry presents the (I, V) relation shown in Figure 14.1b which basically is an ordinary diode characteristic turned up-side down. Taking into account the diode parameters, the mathematical relation between V and I is given as

$$ I = I_{PV,cell} - I_D = I_{PV,cell} - I_{0,cell} \left\{ \exp\left(\frac{qV}{akT} \right) - 1 \right\} \qquad (14.4) $$

where $I_{PV,cell}$ is the current generated by the irradiated light, and I_D is the Schockley diode equation with $I_{0,cell}$ being the leakage current, q the electron charge, k Boltzmann's constant, T the junction temperature in Kelvin, and a the diode ideality constant [5]. A practical PV-array is constructed by stacking a number of PV-cells in either a serial or parallel manner. This will account for additional parameters to the basic equation

$$ I = I_{PV} - I_0 \left\{ \exp\left(\frac{V + R_S I}{V_t a} \right) - 1 \right\} - \frac{V + R_S I}{R_P} \qquad (14.5) $$

where $V_t = N_s kT/q$ represents the thermal voltage of N_s serially connected cells, and R_S and R_P are equivalent series and parallel resistances as illustrated in Figure 14.2a. The corresponding (I, V) curve of this relatively simple but still accurate single-diode model is shown in Figure 14.2b which also indicates the important Maximum Power Point (MPP), being a parameter normally provided by the manufacturer and which is essential for operating the PV-array optimally.

14.4 Rechargeable Batteries – Technologies and Models

Although technologies for rechargeable batteries have not undergone the same extraordinary evolution as compared to transistor technologies, it is however a fact that general improvements have been achieved since the

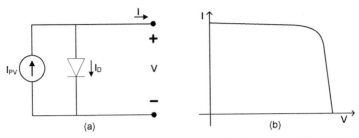

Figure 14.1 (a) Circuit equivalent of a PV-cell, and (b) the corresponding (I, V) curve.

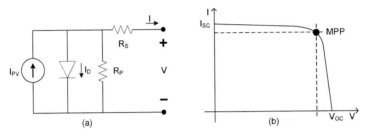

Figure 14.2 (a) Model of a practical PV-array, and (b) the corresponding (I, V) curve including the Maximum Power Point (MPP) as well as the short-circuit current, I_{SC}, and the open-circuit voltage, V_{OC}.

1950s where the Nickel-Cadmium (NiCd) technology was first commercially deployed [6]. Being now a mature technology, NiCd has over the years successfully been applied in various portable electronic devices. The good news about the technology is that is features low cost, and high discharge rates, but at the same time it also has a rather low energy density, and even worse, an inadvertently negative environmental impact upon disposal. The early 1990s brought forward a significant improvement in terms of the Nickel Metal-Hybrid (NiMH) technology showing almost twice the energy density as compared to NiCd. On the down-side however, NiMH batteries have shorter cycle life, are more expensive, and are unable to provide the same high discharge rate as NiCd.

Shortly after the commercial introduction of the NiMH technology, yet another new technology was announced; the Lithium Ion (Li-ion) battery. It provides a significantly higher energy density, and a cycle life about twice that of NiMH. The first generation Li-ion batteries was known to be somewhat chemically unstable when improperly used, but their long lifetime has made them the favourable choice for many portable devices. One of today's emerging types of batteries is the Lithium Polymer (Li-Pol) technology which is remarkable in its physical design. Possibilities for ultra thin design and light weight combined with improved energy density and safety (as compared to the Li-ion technology) are promising characteristics which enables this technology to take major market shares for portable devices. Some of the problems still pending with Li-Pol are manufacturing expenses and challenges with internal thermal management.

No matter which technology we are opting for, two parameters are of special interest: (i) the battery cycle life, and (ii) the battery lifetime. As per definition, the battery cycle life is the number of complete charge/discharge cycles a battery can conduct (typically 500–1200) until its nominal capacity is degraded to below 80% of its initial capacity. In some cases, cycle life may also be defined as the number of cycles the battery can perform until its internal resistance exceeds a predetermined amount, e.g., double its initial value. The cycle life depends on the Depth of Discharge (DoD), and assumes that the battery is fully charged and discharged in each cycle. However, for most cell chemistries, if the battery is only partially discharged each cycle then the cycle life number will be much higher. It is therefore important that we care about the DoD parameter, in the design phase as well as in practical operation, because the shallower the discharge, the more cycles the battery will provide. The exception to this rule exists for NiCd batteries (and to a lesser extent for NiMH) where a partial discharge scheme gives rise to the so called memory effect which can only be reversed by deep discharging. Therefore, the more intelligent a discharge profile we can design, the longer the battery can be employed operational.

Optimizing the other mentioned design parameter, the battery lifetime, basically is a matter of applying theories and methods for lowering the energy consumption at the source, in our case the DSP algorithms executing on the multiprocessor architecture, and at the same time (or eventually as a separate alternative strategy) to take advantage of the fact that the battery being a non-linear device which lifetime depends on the discharging profile. The energy which can be delivered from the battery, being upper bounded by the *theoretical capacity*, depends on the discharge current profile. Using a standard load condition, normally specified by the manufacturer in terms of a constant discharge current at a specified temperature, the energy which can be provided by the battery is know as the *standard capacity*. Finally, the *actual capacity* of the battery is the amount of energy which can be provided under a given load, and it indicates how "battery efficient" a given load system is. A battery-efficient load is characterized as a system which exhibits a discharge profile that enhances the actual capacity. Note that the actual capacity can possibly exceed the standard capacity but it will always be lower than or at best equivalent to the theoretical capacity. From these definitions it is obvious that the battery life can be extended by controlling the discharge current level and shape. When a battery is continuously discharged, a high discharge current will cause it to provide less actual energy until the end of its lifetime as compared to a lower discharge current. This effect is known as the Rate Capacity effect. Similarly, during periods of low or no discharge current, charge recovery takes place meaning that the battery will regain some capacity. This phenomenon is known as the Recovery Effect.

Although the battery in many cases is being considered a trivial, simple, and ideal component which can deliver a given current at a certain voltage for a specified amount of time, it goes without saying that in a realistic design scenario there is more to it. In order not to be too detailed in a design scenario, we normally would rely on a model, and in terms of the discharging scheme, we will therefore apply battery modeling which typically belongs to one of four different classes: (i) analytical models, (ii) electrical circuit models, (iii) stochastic models, and (iv) electro-chemical models. Being beyond the scope of this chapter to discuss exhaustively the various models, we will provide some overall comments only. Analytical models provide values on battery capacity and lifetime based on parameters such as discharge current, operating environments, as well as physical properties of the battery. Normally, such models are computationally simple but at the same time also relatively limited in the discharge effects they can handle, e.g., the recovery effect is included in the most advanced models only. Being typically based on SPICE models, the electrical circuit models provides the possibility to describe accurately the behaviour of many different battery technologies taking into account the rate capacity effect, but fails to account for the recovery effect. In contrast, the stochastic models, which describe the battery behaviour in terms of a discrete-time stochastic process modelling a finite number of charge units, are able to capture the rate capacity effect as well as the recovery effect. Finally, since the electro-chemical models are tailored to specific battery technologies they are very accurate, but similarly also inflexible and computationally intractable.

Even though no single battery model type provides all the benefits in terms of accuracy, multi-metric monitoring, and low computational complexity, analytical models are often opted for in embedded systems design. Among the most popular is the model suggested by Rakhmatov et al. [7], which for a time-varying discharge current $i(t)$ is of the following form:

$$\alpha = \int_0^L i(\tau)d\tau + 2\sum_{m=1}^\infty \int_0^L i(\tau)e^{-\beta^2 m^2(L-\tau)}d\tau \qquad (14.6)$$

where L is the lifetime, and α and β are battery specific parameters. In a constant discharge current scenario i.e., $i(t) = I$, the model reduces to

$$\alpha = I\left[L + 2\sum_{m=1}^\infty \frac{1 - e^{-\beta^2 m^2(L)}}{\beta^2 m^2}\right] \qquad (14.7)$$

The constant α represents the actual battery capacity in terms of charge units (Coulombs), and β^2 captures its non-linear discharge profile, i.e., a small β-value means that for the battery to recover, the load must be reduced. Similarly, a high β-value indicates a "good" battery – with β being sufficiently large, it is seen that Equation (14.7) turns into the model for an ideal power source, $\alpha = IL$.

14.5 Energy Harvesting Considerations

Having now elaborated on appropriate models for PV-arrays and rechargeable batteries, we next prepare for discussions on multiprocessor scheduling in EH-driven environments by considering how to handle simultaneously PV-EH and rechargeable batteries in a combined energy-harvesting and -storage system. First, we note that for wireless embedded applications, the battery cycle life as well as the battery lifetime are important parameters to optimize. For a PV-EH scenario, the good news is that we potentially are able to top-up charge on the battery whenever our portable device is exposed by light. Given that the amount of harvested energy is comparable to the energy consumed by the source, this will account for a limited DoD and thus a cycle life beyond the nominal value. For out-door sensor motes transmitting only few and short messages this may be a realistic assumption, whereas e.g., PV-equipped wireless handsets, which are not very often exposed to full sunlight, will have a higher DoD.

In order to cope with both scenarios, we therefore assume that the battery is being charged through a loss-less regulator when (i) $DoD > 0\%$, and (ii) there is light enough to bring the PV-array output voltage above the battery nominal voltage (in practice the regulator efficiency is less than 100% but state-of-the-art circuitry can bring it close). From an ideal point of view, the

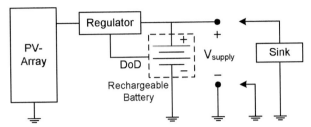

Figure 14.3 EH is conducted by a PV-array which is feeding the sink (eventually including a DC-DC converter) and charging the battery through a charging regulator which is monitoring DoD.

regulator should control the charging current such that for the given PV-array and battery technologies, the (I, V) curve settles at or near by MPP in order to gain a high utilization. Characteristic PV-values such as V_{OC}, I_{SC}, and the MPP voltage and current are usually provided by the manufacturer, and these values are required in order to establish a high-performance regulator model. Research in this direction has been published recently [8].

Once the battery's nominal voltage is reached, the regulator shut-down the current in order to prevent over-charging, and the charging cycle terminates. Figure 14.3 shows the overall energy-harvesting and storage system architecture where there is no feedback from the sink to the regulator. The reason why we consider the sink being an autonomous unit (including an appropriate DC-DC converter) is basically due to the fact that we are addressing static scheduling of the individual algorithmic tasks, and thus no computing overhead is needed to control the execution on the architecture.

In most cases, PV-EH is considered being a highly stochastic process – out-door as well as in-door – because the light intensity will vary according to day/night, local weather/climate conditions, and human related circumstances such as electric light being switched on and off. In the worst case scenario the battery will totally deplete ($DoD = 100\%$), and the system will fail. In order to alleviate such a situation, interesting work on reliable solar prediction algorithms has been suggested, see e.g. [9], the drawback being that such algorithms account for a computational and thus energy consuming overhead. As a first approach, in order to enable a fair and reliable description of the harvesting process we suggest instead that (i) a lower limit on the harvesting/non-harvesting duty cycle is known, and (ii) that the PV-generated power at the harvesting period is also known and constant. Employing according values, we can estimate E_H, denoting the minimum amount of energy which can be harvested during a harvesting period, where

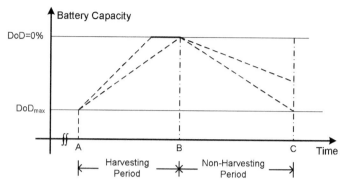

Figure 14.4 The charge/discharge cycle must be kept between $DoD = 0\%$ and DoD_{max}.

this energy is used to (i) charge the battery (from its actual charge level which is somewhere between $DoD = 0\%$ and DoD_{max}, a value we want to specify due to an eventually given lower limit on the cycle life), and (ii) execute the DSP algorithm (on one or multiple interconnected processors), i.e.,

$$E_H = E_{Batt} + E_{DSP,AB} \qquad (14.8)$$

This is illustrated in Figure 14.4 which shows a charge/discharge scenario where the area below the dashed lines indicates possible battery capacities during the charge/discharge cycle (here inaccurately indicated as linear functions). At time $t = A$, a new harvesting period (which terminates at $t = B$ and for which we assume that we know the relative length) is initiated, and we assume, according to Equation (14.8), and our previous set of arguments, that we can harvest sufficient energy to charge the battery while also supplying the executing DSP algorithm. With this assumption, the battery will be charged to at most $DoD = 0\%$ at the latest at $t = B$. If $DoD = 0\%$ is achieved before $t = B$, the regulator will maintain the battery fully charged (eventually with hysteresis), and at the same time supply the processor(s). Similarly, at time $t = B$ the discharging, i.e., non-harvesting period is initiated, and at time $t = C$, the battery is discharged to DoD_{max}, at most. The energy consumed during the non-harvesting period is defined as E_{NH}, and since we (most ideally) cannot allow the overall energy consumption to be larger than the energy production, then

$$E_H - E_{DSP,AB} \geq E_{NH} \qquad (14.9)$$

and thus

$$E_{Batt} \geq E_{DSP,BC} \qquad (14.10)$$

which specifies a lower bound on the energy to be accumulated during the harvesting period. Now, a very interesting observation is that the sample period T which specifies an upper bound on the processing time for one iteration of the DSP algorithm, normally would be much shorter than the harvesting as well as the non-harvesting period, and consequently we have to find means of energy optimization being independent of T. One such technique is to adjust the number of processors to be allocated in the target architecture.

14.6 How Many Processors?

From the above assumptions and discussion it is now clear that our next task is to determine a multiprocessor schedule which complies with two requirements. First of all, the DSP computations are upper-bound time-constrained by the sample interval T. Consequently, we have to design a scheduling functions g, as defined by Equations (14.1), (14.2), and (14.3), and next apply g in order to generate a schedule s with $makespan = \sum_i w_i$ shorter than or equal to T. However, except for the hard to obtain 100% processor-utilization scenario, the processors encounter idle time due to IPC, and the makespan is therefore realistically defined as

$$makespan = \max_{p_k \in Sp} \left\{ \sum_{n_a \in \bar{g}(p_k)} w_{n_a} + t_{idle}(p_k) \right\} \qquad (14.11)$$

Secondly, the generated multiprocessor schedule (where a single processor configuration is simply considered as a special case of a general multiprocessor architecture) has to be designed such that the energy consumption during the non-harvesting period do not exceed E_{Batt}. Both requirements can be accomplished or are subject to optimization by varying the number of processors in the architecture. Normally, by utilizing the inherent algorithmic parallelism, the makespan requirement can be fulfilled by allocating more processing units into the architecture. If the time constraint is initially violated because of too few processors, i.e., too large algorithmic grain sizes, we could opt for adding yet another processor to the architecture and then redo the scheduling. One should note, however, that continuous allocation of more and more CPUs will not necessarily cure the timing problem if it remains. The reason being that an increase in the number of processors naturally will lead to smaller grain sizes in order to exploit the inherent parallelism, which in itself generates more IPC and therefore counteracts the higher speedup gained by having more processors. This is also known as the saturation effect. Alternatively, one could possibly choose another scheduling heuristic which better suits the problem, e.g., in terms of algorithmic precedence structure, number of processors, and computation/communication ratio.

At the same time, however, the allocation of more processors may lead to a reduction in the energy consumption which is contingent on fundamental characteristics of the silicon technology [10]. For CMOS, the power consumption, P, complies with the following relationship:

$$P \propto (V_{dd} - V_t)^2 \qquad (14.12)$$

where V_{dd} is the supply voltage, and V_t denotes the threshold voltage of the given CMOS technology. Decreasing V_{dd} obviously would lead to a square related reduction in the power consumption, the drawback being that the circuit speed, S, will fall according to

$$S \propto (V_{dd} - V_t)^2 / V_{dd} \qquad (14.13)$$

where normally $V_t \ll V_{dd}$. This means that $P \propto V_{dd}^2$, and $S \propto V_{dd}$, thus leading to $P \propto S^2$. Now, let us assume that we have a mono-processor architecture on which we want to obtain a computing capacity of c. This requires a power consumption equal to $k \cdot c^2$, where k is a constant. On the other hand, we may want to obtain the same computing capacity on a parallel architecture with p processors, each having capacity c/p. In that case, the power consumption would be $p \cdot k \cdot (c/p)^2$ which tells us that the power consumption on a p-processor architecture is

$$P(p) \propto \frac{k \cdot c^2}{p} \qquad (14.14)$$

which should be compared to the power consumed by a mono-processor with the same computing capacity.

Interestingly, Equation (14.14) shows that the power consumption is approaching zero as $p \to \infty$, provided that we gradually lower the supply voltage. Unfortunately, the saturation effect mentioned above, contradicts this most wanted (but also ideal) situation, and therefore we have to search for an optimal number of processors lower bounded by our constraints on T as well as on E_{Batt}. Which one is the dominant requirement depends on the actual processors applied in the architecture, the interconnection topology, and the grain clustering. Notice that the energy related IPC overhead has not been included in these considerations which is a pending need for providing more energy-realistic processor-count estimations.

14.7 Cooking It All Together

So far the discussion has been focused on the various components needed in a realistic and efficient PV-aware multi-processor scheduling environment. For each of these items we have highlighted pros and cons, and we have pointed out several open questions still remaining in order to generate accurate static Time- and Energy-optimal (or -constrained) multiprocessor schedules in a computationally tractable manner. In the following, we will emphasize some other research challenges ahead asking for a solution in order to fully integrate the components for streamlined operation.

The scheduling function g basically has the following purposes; it should calculate (i) task assignment onto the processors, (ii) task ordering and IPC allocation, and (iii) invocation time for every node n_i, subject to maintenance of precedence relations and optimization of an objective function C. Although C can be chosen freely according to given design specifications, the above discussion on mutual time- and energy optimization has shown that it should be a fixed and combined function of two variable, $C = f(makespan, E_{Batt})$. Some considerations are surely needed for designing such a multi-objective cost function.

In many cases, designers are aiming for a makespan that is as short as possible, thus enabling more computational functionalities onto the allocated processors. On the other hand, as we have already argued, there is no obvious reason for squeezing the overall execution time below T. Concerning the energy consumption, we are upper bounded by E_{Batt}, and therefore the combined time- and energy cost function C must comply with

$$minimize\{C\} \Rightarrow (makespan \leq T, E \leq E_{Batt}) \qquad (14.15)$$

Since multiprocessor scheduling in general is an NP-complete optimization problem, no analytical solution for minimizing C is at hand. We therefore have to device heuristic approaches which can help us through the design space exploration and lead to solutions having acceptable makespan and energy figures. First, let us assume a mono-processor architecture onto which we want to map the algorithm G. For each node n_i, we can now estimate the energy consumptions E_i (given the supply voltage V_{dd}), and thus $E_{mono} = \sum_i E_i$. Note that E_{mono} is independent of the schedule s. Stepping up the number of processors, G will now be distributed onto several CPUs, and therefore we have to account for IPC, both in terms of time and energy. Consequently, in a given multiprocessor schedule s, the energy consumption is calculated as

$$E_{multi}(s) = \sum_i E_i + \sum_{(n_j,n_k) \in S_a} E_{IPC(j,k)} \qquad (14.16)$$

$$= E_{mono} + E_{IPC}(s) \qquad (14.17)$$

Equation (14.17) is valid in a homogeneous multiprocessor configuration composed of CPUs equivalent to the CPU used in the mono-processor architecture. Beside, the equation only holds as long as the supply voltage remains unchanged while stepping up from the mono- to a multiprocessor configuration which, according to the discussion above, is not exactly what we want to do. We therefore modify the equation to take into account the varying supply voltage V_{dd} which we assume is applied on all CPUs in the configuration

$$E_{multi}(s, V_{dd}) = \sum_i E_i(V_{dd}) + E_{IPC}(s, V_{dd}) \qquad (14.18)$$

According to our defined cost function as shown in Equation (14.15), the schedule s generated by g now has to comply with the following constraint:

$$\sum_i E_i(V_{dd}) + E_{IPC}(s, V_{dd}) \leq E_{Batt} \qquad (14.19)$$

Two interesting aspects should be noticed. First of all, every single processor p_k may potentially run at its privately defined supply voltage, the drawback, of course, being a more complex hardware architecture with multiple supply voltages needed, and secondly, the target architecture could possibly host heterogeneous processors which potentially will have different computing capacities for the same/different supply voltage(s). Under such circumstances, the energy consumption should be estimated as

$$E_{multi}(s, V_{dd}(p_{j,k})) = \sum_{i,k} E_i(s, V_{dd}(p_k)) + E_{IPC}(s, V_{dd}(p_{j,k})) \quad (14.20)$$

where $V_{dd}(p_{j,k})$ indicates that two processors j and k being interconnected for communication may operate at different supply voltages. This multivariable expression for the energy consumption next has to be integrated (i) either directly with the scheduling function g in order to minimize E during the scheduling procedure, or (ii) as part of an evaluation task which afterwards estimates E for the schedule s generated.

14.8 Final Remarks

In this chapter we have identified some of the challenges associated with static multiprocessor scheduling of DSP algorithms executing in battery- and PV-EH powered embedded wireless systems. We have highlighted many routes that need to be researched before we have a complete photovoltaic- aware multiprocessor scheduling framework, and in particular we have ar- gued that existing scheduling methods, heuristic and meta-heuristic, must be extended with a more sophisticated cost function with energy specific metrics for computation as well as for communication in order to cope with simul- taneous time and energy optimization. We have suggested the foundation for a new multi-objective cost function which can be used for experimentation with various scheduling mechanisms in order to comply with given T and E constraints. Similarly, such a multi-objective cost function is needed to determine the appropriate number of processors to be allocated in the target architecture – a number which is lower bounded by the sample period and the amount of energy that can be harvested and stored on the rechargeable battery.

Combining static multiprocessor scheduling with energy-aware design and photovoltaic-based configurations reveals a very wide research field which potentially includes many and diverse disciplines. In conclusion it is therefore also difficult to state explicitly where to start and which topics to favour. One the other hand, we have no doubt that energy harvesting, and in particular solar based EH, has a bright future for wireless embedded systems, and with the inevitable growing number of hardwired and programmable CPUs which will be embedded in portable wireless devices, the need for formal and structured design methods is strong, so the sooner we launch the research effort, the better.

References

[1] ITRS, *International Technology Roadmap for Semiconductors, 2007 Edition, Systems Drivers*, `http://www.itrs.net/Links/2007ITRS/2007_Chapters/2007_SystemDrivers.pdf`, 2007.

[2] Nvidia Homepage, `http://www.nvidia.com/page/home.html`.

[3] Lee, E. and Messerschmitt, D., Static scheduling of synchronous data flow programmes for digital signal processing, *IEEE Trans. Computers* **36**(1), 24–35, January 1987.

[4] Koch, P., Strategies for realistic and efficient static scheduling of data independent algorithms onto multiple digital signal processors, PhD Dissertation, Aalborg University, 1995.

[5] Villalva, M.C., Gazoli, J.R. and Filho, E.R., Comprehensive approach to modeling and simulation of photovoltaic arrays, *IEEE Trans. Power Electronics* **24**(5), 1198–1208, May 2009.

[6] Lahiri, K., Raghunathan, A., Dey, S. and Panigrahi, D., Battery-driven system design: A new frontier in low power design. In *Proc. 15th Int. Conf. on VLSI Design*, 2002.

[7] Rakhmatov, D., Vrudhula, S. and Wallach, D., A model for battery life-time analysis for organizing applications on a pocket computer, *IEEE Trans. VLSI Syst.* **11**(6), 1019–1030, December 2003.

[8] Barca, G., Moschetto, A., Sapuppo, C., Tina, G.M., Giusto, R. and Grasso, A.D., A novel MPPT charge regulator for a photovoltaic stand-alone telecommunication system. In *Proc. Int. Symp. Power Electronics, Electrical Drives, Automation and Motion*, pp. 235–238, June 2008.

[9] Piorno, J.R., Bergonzini, C., Atienza, D. and Rosing, T.S., Prediction and management in energy harvested wireless sensor nodes. In *Proc. 1st Int. IEEE Conf. on Wireless Vitae*, pp. 6–10, May 2009.

[10] Chandrakasan, A.P., Sheng, S. and Brodersen, R.W., Low-power CMOS digital design. *IEEE Journal of Solid-State Circuits* **27**(4), 473–484, April 1992.

15

Quantum Information Technology for Power Minimum Info-Communications

Masahide Sasaki, Atsushi Waseda, Masahiro Takeoka,
Mikio Fujiwara and Hidema Tanaka

National Institute of Information and Communications Technology, Tokyo, Japan

Abstract

We discuss a long-term solution for realizing the power minimum ICT, revisiting the fundamental laws of physics. In particular, we focus on a new paradigm of quantum information and communications technology (Q-ICT), in the context of reducing the power consumption in optical fiber communications. Our analysis shows that quantum communications will be able to reduce the transmission power by more than 40 dB from current technology. The key is a technology of quantum collective decoding at the receiver side.

Keywords: Capacity, channel matrix, coherent communication, coherent state, deep space optical communications, fiber fuse, Helstom receiver, homodyne detection, mutual information, optical communication, quadrature amplitude and phase, quantum computing, quantum collective decoding, quantum probability amplitude, Shannon formula, quantum noise.

15.1 Introduction

One of the important roles of Green ICT is to reduce energy consumption in travel and logistics by realizing virtual meetings with realistic sensation. It would require high capacity transmission of a variety of contents. In optical fiber communications, however, one faces a serious bottleneck in increasing

R. Prasad et al. (Eds.), Towards Green ICT, 229–243.

the capacity. It is the power limit of a fiber, namely, fiber fuse. When orders of several Watts of optical power is input into an optical fiber, the edge face of a fiber may be burn out. Actually this is the case of intercontinental links through submarine-fiber cables where dense wavelength division multiplexing (DWDM) technology and the power feed for more than 100 fiber amplifiers are necessary for a long-haul transmission with a higher capacity. The data traffic is doubling every year due to the rapid spread of broadband access. It is most likely that the throughput per fiber will reach a Peta (10^{15}) bps level in a few decades. There cannot be seen straightforward solutions by current optical communications technology to sustain this increasing demand.

Network nodes also consume huge power. In particular, high-end electrical routers at a Tbps level consume a few tens kW. A straightforward scaling predicts the power consumption will reach Mega W, which corresponds to a level of electric power plant.

In order to circumvent this bottleneck, various approaches are taken, including new multi-core fibers and fiber-fuse protection devices. In a long-term perspective, however, it is strongly desired to investigate a totally new paradigm which is based on quantum information and communications technology to realize the ultimate resource-efficiency. It is nothing but to study a fundamental issue of communications, namely how to transmit the maximum information through an optical channel with a given finite amount of signal power and bandwidth.

This issue would also be an important concern in a next generation supercomputer in which electrical interconnections will replaced with optical ones for much faster operation. Then data exchange between CPU and memory is nothing but optical communications. As the processing speed increases, one will face the same kind of power density limit as in optical fiber communications. It is most likely that computer architecture should eventually be designed based on optical communications theory.

In this chapter, we first review recent advances of optical communications and forecast near future optical communications. We secondly analyze the transmission rate of a fiber attainable by ideal coherent communications. We then present the quantum limit for a fiber attained by the best strategy allowed by quantum mechanics. We finally show the ultimate bound for a lossy channel using all electromagnetic field modes.

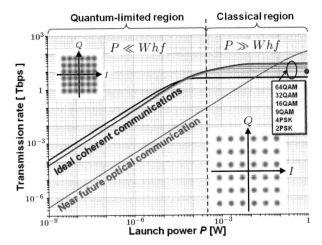

Figure 15.1 Rough estimate of the transmission rate versus the launch power for current, near future optical, and ideal coherent communications. The circle on the vertical line at the right end roughly corresponds to the experiment by Alcatel-Lucent.

15.2 Optical Communication and Beyond

In the laboratory the transmission at 16 Tbps of data through a single fiber over a distance of 2,550 km has been demonstrated by Alcatel-Lucent recently [1]. In this demonstration a 100 Gbps channel modulation per wavelength and a DWDM of about 164 wavelengths are used. The signals are amplified in every 51 km, and totally 50 amplifiers were employed. The launch power, i.e. the input power into the transmitter side of a fiber, is estimated to be roughly 1 W. Figure 15.1 shows a rough estimate of the transmission rate versus the launch power for several kinds of technologies. The right end of the horizontal axis is a point of 1 W, which roughly corresponds to the fiber fuse threshold. The circle on the vertical line at the right end roughly corresponds to the experiment by Alcatel-Lucent.

15.2.1 Near Future Optical Communications

We would like to forecast a performance of near future optical communication based on this experiment, from the viewpoint of how the channel capacity of a commercial fiber scales as a function of the launch power. It is, however, a much involved task to derive the capacity of a practical system, including various components such as a fiber, modulators, amplifiers, switches, and detectors. We should make some assumptions to simplify a channel model to find order estimates.

So we first consider an additive white Gaussian-noise channel with a power and bandwidth constraint. The channel capacity is given by the Shannon formula

$$C_S = W \log \left(1 + \frac{S}{N_0 W} \right) \tag{15.1}$$

where W is a bandwidth of the channel, S is an average signal power, N_0 is a spectral density of Gaussian noise, and $N_0 W$ is the average noise power for the whole bandwidth. We assume that the effective transmissivity of the signals from the fiber input to the output via the amplifiers is roughly $\eta \sim 10^{-2}$. Let the launch power be P. Then $S = \eta P$. The excess noises in the channel and the detectors are assumed to be modeled as a parameter $N_0 W \sim 100 \, \mu W$. The transmission distance does not appear explicitly but is implicitly included in η and N_0. In this way we can roughly model typical characteristics of the state-of-the-art transmission system by Alcatel-Lucent as shown by the circle in Figure 15.1, roughly corresponding to a signal-to-noise ratio (SNR) of 20 dB for the launch power of $P = 1$ W.

The Alcatel-Lucent experiment employed the c-band of 2 THz. A commercial silica fiber, however, has a bandwidth of $W = 50$ THz, corresponding to the wavelength range of about 1.2–1.6 μm. The bottom curve in Figure 15.1 shows the Shannon capacity C_S estimated for the full bands of a fiber. This assumes an advanced technology of multi-ary modulation of light-wave signals and phase-sensitive detection of the signals. This is a performance of a near future optical communication for a single fiber.

This level of technology still suffers from amplified spontaneous emission noise in an amplifier, and the excess noise in a detector. To reduce these excess noises, we have to realize quantum optimal amplifiers and excess-noise free homodyne detectors. We could then expect ideal coherent communications where only the quantum noises dominate. The quantum noise is associated with an optical signal itself which is due to the uncertainty principle of quantum mechanics. Actually the position and the momentum of a particle cannot be measured precisely. The variances of the observed values obey an inequality $\Delta x \Delta p \geq h = 6.6 \times 10^{-34}$ J·s, where h is the Plank constant. In the case of optical field, x and p are the quadrature amplitude and phase.

Figure 15.2 Contour plot of closely packed 4 phase-shift keyed coherent signals in the phase space.

A state of a signal pulse is then represented by a pure coherent state

$$|\alpha\rangle = \exp\left(-\frac{|\alpha|^2}{2}\right) \sum_{n=0}^{\infty} \frac{\alpha^n}{\sqrt{n!}} |n\rangle, \qquad (15.2)$$

where $|n\rangle$ is an energy (photon number) eigenstate defined for each pulse. The $|\cdot\rangle$ is the Dirac notation of quantum state, called a ket vector, and α is a complex amplitude of an optical field mode. The signal encoded into the coherent state is usually detected by homodyne detection where the signal is first combined with a strong local oscillator field, and is then converted into a photo-current output signal. In this way one can measure the real and imaginary part of the complex amplitude $\alpha_R (= \text{Re}[\alpha])$ and $\alpha_I (= \text{Im}[\alpha])$, called quadrature amplitude and phase. The conditional probability of having an outcome (y_c, y_s) for an input α is given by

$$p(y_c, y_s | \alpha_R, \alpha_I) = \frac{1}{\pi} \exp\left[-(y_c - \alpha_R)^2 - (y_s - \alpha_I)^2\right]. \qquad (15.3)$$

The variances of x_c and x_s is the manifestation of the quantum noise Δx and Δp, which can never be eliminated, and hence eventually limits the communication performance.

A contour plot of the probability distributions for the case of quaternary phase shift keying is depicted in Figure 15.2. The blurred circles schematically represent quantum noise variances. When the average photon number $|\alpha|^2$ of the signal $|\alpha\rangle$ is small, the quaternary signals are closely packed, and suffers from an inevitable error due to the quantum noise.

15.2.2 Ideal Coherent Communications

We now consider WDM schemes to find the achievable transmission rate by the ideal coherent communications with quantum optimal amplifiers and dyne-type detections. We divide the full bandwidth $W = 50\,\mathrm{THz}$ of a fiber equally into $N_C = 400$ channels. Due to the Fourier transform limit, we choose the pulse generation rate r as $r = 9.95 \times 10^9$ pulse/s/channel which results the total number of signal pulse propagating through the fiber is $r \cdot n = 3.98 \times 10^{11}$ pulse/sec.

In each channel we modulate the coherent state pulse into multi-ary phase shift keyed (PSK) or quadrature amplitude modulation (QAM) formats, $\{|\alpha_1\rangle, |\alpha_2\rangle, \cdots, |\alpha_M\rangle\}$ with the equal prior probability $1/M$. An average photon number $\bar{n}$ is given by

$$\bar{n} = \frac{1}{M} \sum_{i=1}^{M} |\alpha_i|^2. \tag{15.4}$$

The total input power of our model is thus given by

$$P = r \cdot \frac{1}{M} \sum_{i=1}^{M} |\alpha_i|^2 h \sum_{j=1}^{n} f_j \quad [\mathrm{W}], \tag{15.5}$$

where f_j is the center frequency of j-th channel.

The amount of transmissible information per pulse in this channel is estimated by Shannon's mutual information

$$
\begin{aligned}
&I(X; Y) \\
&= \sum_x P(x) \sum_y P(y|x) \log \left[\frac{P(y|x)}{\sum_{x'} P(x')P(y|x')} \right],
\end{aligned} \tag{15.6}
$$

where $P(y|x)$ is the channel matrix given by Eq. (15.3). The total transmission rate in the WDM channel is given by

$$T = n \cdot r \cdot I(X; Y) \quad [\mathrm{bit/sec}]. \tag{15.7}$$

The performances of ideal coherent communications for six cases of BPSK, 4PSK, 9QAM, 16QAM, 32QAM and 64QAM signals are shown in Figure 15.1. We can recognize two characteristic regions; the flat region and the linear region. This flat region is the classical region, where the launch power is greater than the total quantum noise power $P \gg Whf$, and the signals are orthogonal to each other. In this region the transmission rate is directly determined by the multiple level of modulation. But from the viewpoint of power saving, it is not optimal. We still waste redundant power.

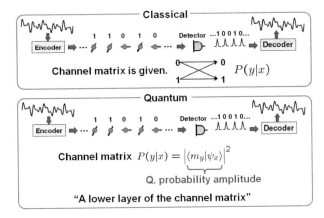

Figure 15.3 Channel matrix and quantum probability amplitude.

In the linear region, on the other hand, the launch power is greater than the total quantum noise power $P \ll Whf$ and hence we spend less power, but then cannot make the signal distant apart, and the quantum noises directly limit the transmission rate. The transmission rates decreases as the input power. This is the quantum-limited region. The subject here is what we can do to improve the performance in this region.

15.3 Quantum Communication

In classical scheme, the channel matrix is given. In the optical communications discussed in the previous, it is given by Equation (15.3). We do not ask further detailed structure of probability. In the quantum scheme, however, the channel matrix is given by the absolute square of quantum probability amplitude. It is an inner product between the measurement vector $|m_y\rangle$ and the signal state vector $|\psi_x\rangle$, as $P(y|x) = |\langle m_y|\psi_x\rangle|^2$ [5]. Thus in the quantum domain, there exists a lower level layer in the channel matrix, and we can directly control this probability amplitude directly to induce an appropriate quantum interference for a better SNR. They are summarized for comparison in Figure 15.3

15.3.1 Quantum Detection

This quantum control is actually made in the detection process. It is categorized into two main classes. One is the separable detection, and the

other is the collective decoding. In the separable detection, the receiver de-
tects each coherent state pulse individually, namely in a single-shot manner.
After the separable detection, One then has the classical signals, and the
sequence of them are decoded a classical way. The well known detection is
known as the Helstrom receiver which attains the minimum bit error rate [5].
Mathematically it is given by a set of positive matrices

$$\hat{m}_y = |m_y\rangle\langle m_y|, \tag{15.8}$$

$$|m_y\rangle = \hat{G}^{1/2}|\sqrt{\eta}\alpha_y\rangle, \tag{15.9}$$

with

$$\hat{G} = \sum_{x=1}^{M} |\sqrt{\eta}\alpha_x\rangle\langle\sqrt{\eta}\alpha_x|, \tag{15.10}$$

satisfying the resolution of identity

$$\sum_{y=1}^{M} \hat{m}_y = \hat{I}, \tag{15.11}$$

This direct control of quantum probability amplitude in the detection process brings a new remarkable effect, when we consider the coding. It is the
super-additive coding gain. This can be summarized in the following way.
When we increases the transmission resources by n times, then the capacity
can increase even more than n times. Classically, however, the capacity increases n times at most, and never more than that. The very origin of this
effect is quantum computing, which is performed prior to the measurement.

In quantum computing, we can make a state, which is in a 0, and also
in a 1 simultaneously. This new notion is called quantum bit, or qubit in
short, and allows us to realize super hyper parallel processing. An example
of length 2 coding is schematically shown in Figure 15.5. We first transform the received pulses into a superposition of several possible sequences
by a quantum computer, and then perform a measurement afterward. In the
measurement process, these probability amplitudes interfere with each other,
reducing the decoding error. This kind of new decoding is called quantum
collective decoding.

In the classical scheme, in contrast, each pulse is first measured sep-
arately, producing all possible sequences. They are then decoded by classical processing. In that case, the channel matrix element is simply the
product of these two. Within this range, we can never reach this level of
distinguishability.

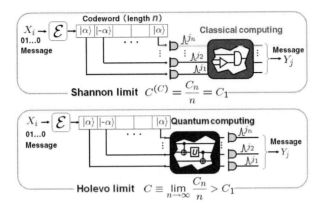

Figure 15.4 Comparison of quantum and classical decoding schemes.

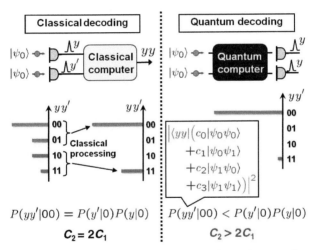

Figure 15.5 Quantum and classical decoding in the case of length two coding. The quantum interference at the probability amplitudes reduces the decoding error.

The important principle of super-additive coding gain was demonstrated three years ago by NICT. In the experiments, single photon states in the polarization-location coding were used [2].

One of well known mathematical construction of quantum collective decoding for a general case of length-n codeword

$$|\Psi_{\mathbf{x}}\rangle \equiv \left|\sqrt{\eta}\alpha_{x_1}\right\rangle \otimes \left|\sqrt{\eta}\alpha_{x_2}\right\rangle \otimes \cdots \otimes \left|\sqrt{\eta}\alpha_{x_n}\right\rangle. \qquad (15.12)$$

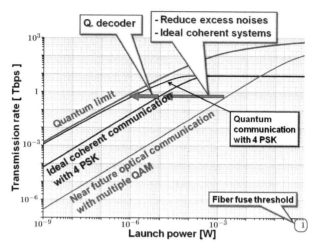

Figure 15.6 How to reach the quantum limit of the single-fiber capacity.

is given by [4, 6, 7]

$$\left| M_\mathbf{y} \right\rangle = \hat{G}^{1/2} \left| \Psi_\mathbf{x} \right\rangle, \tag{15.13}$$

with

$$\hat{G} = \sum_\mathbf{x} \left| \Psi_\mathbf{x} \right\rangle \left\langle \Psi_\mathbf{x} \right|. \tag{15.14}$$

The channel matrix of having an outcome $\mathbf{y} = (y_1, y_2, \cdots, y_n)$ given an input $\mathbf{x} = (x_1, x_2, \cdots, x_n)$ is given by

$$P(\mathbf{y}|\mathbf{x}) = \left| \left\langle M_\mathbf{y} \middle| \Psi_\mathbf{x} \right\rangle \right|^2 \tag{15.15}$$

The decoding described by $\{ \left| M_\mathbf{y} \right\rangle \}$ can be in principle implemented as the lower schematic in Figure 15.4.

When this super-additive coding gain by $\{ \left| M_\mathbf{y} \right\rangle \}$ is used up for a very large block sequences, a much larger capacity can be achieved [4, 6, 7]. In Figure 15.6, the capacity for 4PSK signals obtained by such a quantum collective decoding is shown by the black solid line. The gain in saving the launch power is roughly 10dB from the ideal coherent communication.

15.3.2 Ultimate Capacity of a Lossy Optical Channel

Finally the ultimate capacity is achieved by increasing the modulation level to dense coherent modulation and by using quantum collective decoding. Here we briefly review the recent result of the ultimate capacity.

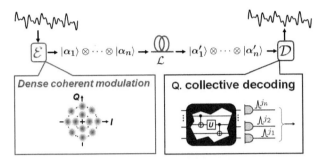

Figure 15.7 Optimal transmission scheme for a lossy optical channel.

We consider a lossy channel of electromagnetic field, often referred to as a lossy bosonic channel. We assume a frequency-independent transmittance η. In our context, given this lossy channel, and the power and bandwidth constraints, one could do anything allowed by quantum mechanics to transmit the maximum information. Thus the ultimate capacity must be derived by fully quantum mechanical optimization of encoding and decoding strategies. The solution of this difficult problem has recently been solved [3]. This tells us that the optimal encoding is given by the dense coherent modulation, which is completely conventional coherent communications technology. The optimal decoding, on the other hand, essentially requires quantum effects. Namely this should consists of quantum computing with coherent states to transform the received codeword state into an appropriate quantum state, and the final measurement on it afterward. This is quantum collective decoding as mentioned above.

An important feature of the coherent state is that its purity can never be lost even after propagating through a lossy channel. The pure state input $|\alpha\rangle$ comes out of the channel as the pure state $|\sqrt{\eta}\alpha\rangle$. No other state satisfies this property. This is the reason why the coherent state modulation is the optimal encoding strategy.

The capacity can be calculated as follows. Let the frequency cutoff be f_L and f_U. The capacity is given by

$$C_Q = \int_{f_L}^{f_U} df g\left(\eta \cdot \bar{n}(f, \beta)\right), \tag{15.16}$$

where $g(x) = (x+1)\log_2(x+1) - x\log_2 x$ and $\bar{n}(f, \beta)$ is the optimal photon number distribution given by

$$\bar{n}(f, \beta) = \frac{1/\eta}{\exp\left(\frac{\beta h f}{\eta}\right) - 1}, \tag{15.17}$$

with a Lagrange multiplier β that is determined through the constraint on the average transmission power

$$P = \int_{f_L}^{f_U} df \, hf \, \bar{n}(f, \beta) \, [\text{W}]. \tag{15.18}$$

This quantum theoretical limit C_Q for the full telecom-fiber bandwidth $W = 50\,\text{THz}$ is shown by the top curve in Figure 15.6. As seen, the capacity C_Q for the current single telecom-fiber can never exceed Peta (10^{15}) bps when the launch power is within the fiber fuse threshold 1W for a channel transmittance of $\eta = 10^{-2}$. Toward this ultimate limit, We should first reduce the excess noises and realize ideal coherent system. We should finally use dense coherent modulation and quantum collective decoding. Eventually we could then get 30 dB gain from the near future optical communications both in the power budget and the transmission rate.

To overcome the fiber-capacity limit, we should find new transmission lines of light. But if we could do this, we will eventually face the very final bound, which limits any communications using electromagnetic fields. This line is shown right here by the top curve in Figure 15.8. This is obtained by taking $f_L, f_U \to \infty$ as

$$C_Q^\infty = \frac{\pi}{\ln 2}\sqrt{\frac{2\eta P}{3h}}. \tag{15.19}$$

Nobody can go beyond this limit no matter how much capacity we want. Thus quantum information theory tells us that the capacity is finite. The very origin of this is the quantum noise.

To attain this capacity one should be able to perform quantum computing on the received codeword over all the relevant modes. We could eventually reduce the power budget by more than 40 dB from the current level of technology. The practical code construction and physical implementation of such quantum collective decoding are completely unknown. Only the existence of such schemes is ensured theoretically.

This kind of quantum decoder will also be indispensable for deep space optical communications. It is very long distance, but there are no amplifiers.

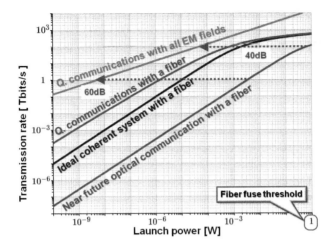

Figure 15.8 The very final bound of communications and the capacity limits of a fiber.

The power feed is very limited. This is a really quantum-limited channel. Quantum theory tells us that, we will be able to extend a Gbps link to Mars from Earth. We have assumed that the bandwidth is $W = 1\,\text{THz}$, 8-channel WDM with the repetition rate of 125 G pulse/s/channel is used, the launch power is 1 W, and the link is made by the optical antenna of 0.3 m for the transmitter and 10 m for the receiver. When one could realize quantum decoder for 1 PHz bandwidth, it is possible to realize Tbps link to Mars. These evaluations are shown in Figure 15.9.

15.4 Conclusion

We studied the ultimate capacity limit of a lossy optical channel. We calculated the capacities for several coherent and quantum communication systems, including the WDM transmission of coherent PSK and QAM signals in a commercial fiber.

The first step task toward the power-minimum ICT is to reduce excess noises in the current optical communication systems, and to develop the quantum optimal amplifier. Then by applying the multi-ary coherent mudulation and the homodyne detection, we could reduce the launch power to attain a given transmission rate, by more than two orders of magnitude from the current level.

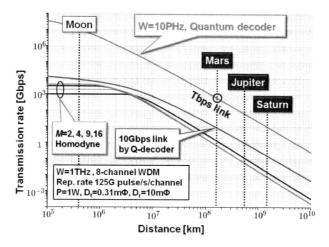

Figure 15.9 Predictions on the capacities for deep space optical communications.

The next step is to develop the quantum collective decoding. This allows one to reduce the launch power to attain a given transmission rate, by more than 40dB from the current level. By increasing the number of multi-ary signals, one can get close to the ultimate quantum limit C_Q. We have seen that the capacity C_Q for the current single telecom-fiber can never exceed Peta (10^{15}) bps when the launch power is 1 W for a channel transmittance of $\eta = 10^{-2}$. It should be noted that although the collective measurement for long-length codewords is still a quite challenging technology at present, the small scale collective decoding can be useful in boosting the performance of optical communications when combined with large scale classical coding [2, 8].

References

[1] Charlet, G., Renaudier, J., Mardoyan, H., Tran, P., Bertran Pardo, O., Verluise, F., Achouche, M., Boutin, A., Blanche, F., Dupuy, J.-Y. and Bigo, S., Transmission of 16.4Tbits/s capacity over 2,550 km using PDM QPSK modulation format and coherent receiver. In *OFC/NFOEC 2008*, San Diego, USA, 2008.

[2] Fujiwara, M., Takeoka, M., Mizuno, J. and Sasaki, M., Exceeding the classical capacity limit in a quantum optical channel. *Phys. Rev. Lett.* **90**(16), 167906, April 2003.

[3] Giovannetti, V., Guha, S., Lloyd, S., Maccone, L., Shapiro, J.H. and Yuen, H.P., Classical capacity of the lossy bosonic channel: The exact solution. *Phys. Rev. Lett.* **92**(2), 027902, January 2004.

[4] Hausladen, P., Jozsa, R., Schumacher, B., Westmoreland, M. and Wootters, W.K., Classical information capacity of a quantum channel. *Phys. Rev. A* **54**(3), 1869–1876, September 1996.

[5] Helstrom, C.W., *Quantum Detection and Estimation Theory*. Academic Press, New York, 1976.

[6] Holevo, A.S., The capacity of the quantum channel with general signal states. *IEEE Transactions on Information Theory* **44**(1), 269–273, 1998.

[7] Schumacher, B. and Westmoreland, M.D., Sending classical information via noisy quantum channels. *Phys. Rev. A* **56**(1), 131–138, July 1997.

[8] Takeoka, M., Fujiwara, M., Mizuno, J. and Sasaki, M., Implementation of generalized quantum measurements: Superadditive quantum coding, accessible information extraction, and classical capacity limit. *Phys. Rev. A* **69**(5), 052329, May 2004.

16

Approaches for Green Communications to Improve Energy Efficiency

Masahiro Umehira

Department of Media and Telecommunications Engineering, College of Engineering, Ibaraki University, Ibaraki 316-8511, Japan

Abstract

This chapter describes how the communications network can be used to improve the energy efficiency of the human life, i.e. energy saving by communications and energy saving of communications. It also discusses the metrics of the energy efficiency of the communications network, and describes the concept of "Sleep Mode Anywhere", and current on-going efforts towards green communications such as Energy Efficient Ethernet.

Keywords: Communications networks, energy efficiency, metric of energy efficiency, W/bit, sleep mode.

16.1 Introduction

After IPCC (Intergovernmental Panel on Climate Change) released the fourth assessment report on climate change in 2007 [1, 2], the environment friendly technologies such as low power consumption consumer electronics products and motor vehicle with high fuel efficiency, i.e. green technologies for a sustainable society, have been drawing a significant deal of attentions. Currently, the Special Report on "Renewable Energy Sources and Climate Change Mitigation" is being prepared to provide a better understanding and broader

R. Prasad et al. (Eds.), Towards Green ICT, 245–264.

information on the mitigation potential of renewable energy sources and it will be released in 2010.

On the other hand, in the area of information and communications technologies, the concept of "computer located anywhere" by Sakamura [3] and "ubiquitous computing" [4] by Weiser have been becoming a reality thanks to recent advances in electronics and telecommunication technologies. In the ubiquitous network, many small but smart devices are widely distributed to help human life. It is expected that such small devices can provide various types of services if they are networked. To explore the promising services based on the concept of ubiquitous network, various projects and trials are being conducted [5,6].

From the viewpoint of green communications technologies, one of the key issues in the ubiquitous network is power consumption reduction of the network devices since many devices are widely deployed. The ubiquitous network using the power efficient devices is expected to be used to improve the energy efficiency of the human society. One of the examples of this approach is a famous "smart grid" [7,8], where electricity distribution network will be controlled to improve its energy efficiency by using communications networks.

In addition, communications network is an essential infrastructure for human life nowadays, not only for traditional voice communications but also for multimedia services including Internet access, e.g. email, file downloading, file transfer and entertainment. As the communications network is useful and the broadband network technologies for cellular network and optical fiber network, i.e. FTTH (Fiber To The Home) are significantly advancing, the volume of the information carried over the communications network is increasing rapidly and is expected to continue to grow. Thus, it is indispensable to improve the energy efficiency of the communications network itself, in addition to improving the energy efficiency of the human life.

With the above-mentioned backgrounds, this chapter reviews how the communications network can be used to improve the energy efficiency of the human life and discusses the metrics of the energy efficiency of the communications network. In addition, some approaches toward green communications are discussed in this chapter.

16.2 Future Trend of Information Traffic and Power Consumption

Though there is no report on the prediction of power consumption required for world-wide telecommunications networks in future, it is easily understood that the power consumption required for telecommunications will continue to increase since the information traffic is continuing to grow. Figure 16.1 shows an example of the expected traffic growth of Internet in Japan [9]. As seen in this figure, the traffic volume of Internet is predicted to grow exponentially. The estimated growth of Internet traffic between year 2025 and year 2006 is 190 times according to the fast growth scenario or 104 times according to the steady growth scenario. This prediction seems reasonable since Internet is widely used in these days and the volume of the contents over Internet is becoming larger and larger according as the telecommunications networks evolve. A huge amount of Internet traffic is carried over telecommunications infrastructure consisting of optical fiber networks and wireless networks. To meet this increasing demand, both wired and wireless networks are evolving toward more bandwidth, e.g. IMT-advanced, FTTH such as 10GE-PON and WDM-PON, and IEEE 802.3ba, i.e. 40 Gbps/100 Gbps Ethernet.

As the information traffic increases, the power consumption of telecommunications networks also increases. Table 16.1 shows an example of estimated growth of power consumption related to ICT equipment [10]. The power consumption growth is categorized to four areas, i.e. communications networks, server in data center, PC and display. As shown in this estimation, the total power consumption related to ICT equipment is estimated to increase three to four times in 2026 compared to that in 2006, and it accounts for 15–20% of the overall electric power consumption in 2006 in Japan [9]. Note that the power consumption of communications networks equipment is expected to increase 13 times. Thus, it is urgently necessary to reduce the power consumption of telecommunications networks equipment, without sacrificing the benefits of telecommunications services.

16.3 Approaches for Green Communications

The green communications is to avoid unnecessary power consumption and to save the energy. In this context, there are two approaches related to communications technologies from the view point of saving energy [11]. This means that the energy should be used only when they are really needed for some benefit using communications networks. One is to save the energy required

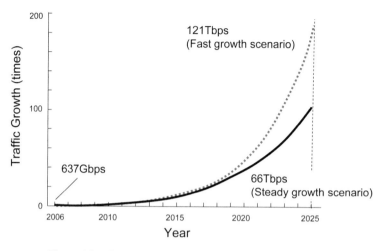

Figure 16.1 Expected traffic growth of Internet in Japan [9].

Table 16.1 Estimated growth of power consumption related to ICT equipment [10].

	Communications Networks	Servers, Data Center	PC	Display
2006	80×10^{12} kWh	214×10^{12} kWh	166×10^{12} kWh	1560×10^{12} kWh
	17%	46%	3.6%	33.4%
2025	1033×10^{12} kWh	527×10^{12} kWh	41.2×10^{12} kWh	816×10^{12} kWh
	43%	22%	1.7%	34%
Growth ratio	13	2.5	2.5	5.2

for human life by using communications networks. Note that we should not sacrifice the comfortable and convenient dairy life with advanced communications services to save the energy. The other is a more direct way to save the energy of communications network itself. Note that the power consumption of the communication equipment should be reduced keeping the quality of communications services.

16.3.1 Energy Saving by Communications

Energy saving by communications includes two approaches. One approach is energy efficiency improvement by using communications, i.e. to save energy by using communications networks. The other approach is to replace human activities with an alternative means using communications networks. However, effectiveness of the latter approach strongly depends on the applications. We need to consider how it will work in real world.

16.3.1.1 Energy Efficiency Improvement by Communications

A typical example of this approach is "smart grid" to control the electricity distribution network by metering the electricity usage using the communications networks in order to use the electricity more efficiently [7]. Essentially, "smart grid" controls electricity supply and demand, and makes it possible to construct distributed electricity generation systems and to store the electricity, by using communications technologies. If the electric energy consumption can be saved by using communications networks, this is "green communications". However, we should note that the benefit of "smart grid" is not only for energy efficiency. As "smart grid" uses digital technologies including communications, it can improve the reliability and the security of the electric system. Therefore, "smart grid" is drawing a great deal of attentions for its various benefits.

Another example similar to "smart grid" is ITS (Intelligent Transport System), which improves the energy efficiency of transportation system using information networks. For example, if traffic jam can be avoided or reduced by controlling and optimizing the car transportation traffic flow, it will be able to save a significant amount of gasoline. Thus, ITS is expected to reduce CO_2 emission caused by driving a car. In addition, the car user can save time in their life. Though ITS is originally aiming at enhanced safety and transportation efficiency, the power saving is becoming more important issue from the viewpoint of green ICT.

16.3.1.2 Replacement by Communications

On the other hand, the benefit of replacing human activities with an alternative means using communications networks is controversial. It strongly depends on the target application whether it is useful to save the energy.

An example is tele-conference. It is said that tele-conference can save the energy for transportation because the conference participants do not consume the energy to go to the conference site from their office. Is this true from the energy saving point of view? It is true that the tele-conference can save time because they do not need to waste their time to go to the conference site. However, if we assume they go to the conference site by public transportation, e.g. train, it should be noted that the energy is consumed to operate the train whichever they take a train or not. Even worse, in addition to the energy for train, the tele-conference equipment also consumes the energy. In this case, tele-conference means "energy for train + energy for tele-conference" from the viewpoint of energy saving. However, if a car is used to go to office, it is a trade-off between the energy to go to the conference site by car and that for tele-conference equipment. The discussion on tele-work is similar to that on tele-conference. This means the tele-work is meaningful from the green ICT perspective if the tele-work results in the scarcer scheduled train, i.e. change in life-style. Without the change of life style, tele-conference and tele-work cannot save the energy, and even worse they can result in the increase of energy consumption.

Another example is "electronic newspaper" which can replace paper based newspaper. In this application, a file of electronic newspaper is delivered to the subscriber via communications networks, while a traditional newspaper is produced by printing a newspaper and is delivered to the subscriber's residence. Therefore, if the energy consumed for newspaper production by paper, delivery via communications networks and browsing it by PC is less than that for newspaper production by paper and delivering it, we can save the energy by using communications networks.

As mentioned above, energy saving by using communications networks is a trade-off issue depending on the applications. Thus, we need to be careful in considering the effectiveness of green communications.

16.3.2 Energy Saving of Communications

The other approach of green communications is to reduce the power consumption of communications networks themselves. There are various approaches to achieve this target, e.g. device technologies, LSI architectures, hardware architecture, sleep mode control, etc., to reduce the energy consumed by communications equipment. In fact, the power saving has been an issue in wireless mobile communications since a mobile terminal is a battery-operated device. However, the power saving of wired communications network equipment has not been seriously considered since AC power is usually available. Power saving must be addressed in both wireless/wired networks to realize green communications anywhere. Technical issues from the view points of energy efficiency include:

1. Energy-efficient protocols and protocol extensions for energy-efficient data transmission.
2. Energy-efficient technology for network equipment such as switch and base station.

Table 16.2 An example of energy efficiency of communications systems.

	ISDN	ADSL	FTTH
Transmission speed (kbps)	64	6500	100000
Environmental load (kg-CO_2/year)	83.2	106.6	57.4
Efficiency (kg-CO_2/kbps)	1.3	0.016	0.00057
Normalized efficiency (W/(bit/s))	1	0.012	0.00043

3. Energy-efficient communications management such as remote power management for terminals.
4. Measurement and profiling of energy consumption.

16.3.3 Metric of Energy Efficiency of Communications Equipment

The figure of merit of communications networks equipment has been addressed in terms of bit/s in communications systems or bit/s/Hz in wireless communications so far. This means the transmission speed of the communications systems and the frequency utilization efficiency in the case of wireless networks. In the case of wired communications, the higher bit/s, the better system. In addition to them, green communications needs another figure of merit to indicate the energy efficiency, i.e. Watt/(bit/s) in addition to bit/s. An example of the figure of merit, Watt/(bit/s) of communications systems is shown in Table 16.2, which compares the energy efficiency of ISDN (Integrated Services Digital Networks), ADSL (Asymmetric Digital Subscriber Loop) and FTTH (Fiber-To-The-Home). Though the environmental load, i.e. power consumption of ADSL is higher than that of ISDN, the energy efficiency of ADSL is better than that of ISDN in terms of W/(bit/s). However, this metric does not seem to indicate the energy efficiency of communications systems because most of the communications equipment wastes the energy while it neither transmits nor receive the information. This must be taken into account to define the metrics of energy efficiency of communications equipment.

As is well known, the sleep mode is employed in cellular networks to lengthen the battery life time of a mobile terminal. Apparently, the sleep mode is useful to improve the energy efficiency and it is indispensable for battery-operated devices. This means it is desirable to make the power consumption of the communications equipment as small as possible while it is not used to carry the information. Considering the above-mentioned, W/bit should be used instead of W/(bit/s) as a metric of the energy efficiency.

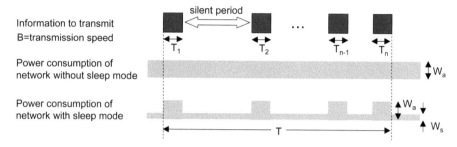

Figure 16.2 Metric of energy efficiency of communications networks equipment.

Figure 16.2 illustrates the metric of energy efficiency of communications equipment. Let us assume B is the bit rate in bit/s. This equipment is active to transmit/receive the information during the period, T_i, where $i = 1, \ldots, n$. If the sleep mode is not supported, the power consumption is constant, i.e. W_a. In this case, the metric in W/bit is equal to W/(bit/s) and is given by

$$\frac{W}{bit/s} = \frac{W_a}{B} \qquad (16.1)$$

On the other hand, if the sleep mode is supported, the power consumption is W_a while it is active and it is W_s when it is in sleep. In this case, the metric in W/bit is given by

$$\frac{W}{bit} = \frac{W_a \sum_{i=1}^{n} T_i + W_s \left(1 - \sum_{i=1}^{n} T_i\right)}{B \sum_{i=1}^{n} T_i} \qquad (16.2)$$

Power consumption of the communications networks equipment/device should be in proportion to traffic volume, i.e. number of bits. To meet this requirement, drastic change is needed in the design of communications network/equipment. The key technique is "sleep mode" which is indispensable in the cellular phone. To support the sleep mode, it needs new design approach in circuit architecture level as well as protocol design level.

16.4 Sleep Mode in Communications Networks

16.4.1 Sleep Mode in Cellular Networks

16.4.1.1 General Sleep Mode Mechanism

In cellular networks, battery life time of the mobile terminal is one of the most critical factors. Therefore, sleep mode is specified to lengthen the battery life

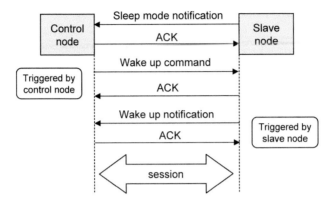

Figure 16.3 Sleep mode control mechanism in cellular networks.

time of the mobile terminal while the mobile terminal is not communicating with the base station. Energy saving in the sleeping mode is supported by the cooperation of the network and the mobile terminal.

Sleep mode of the mobile terminal is an inherent function in cellular systems since the mobile terminal is battery operated. The sleep mode is used to save the battery power when the mobile terminal is waiting to be paged by the control node. In cellular networks, there are two types of nodes to implement the sleep mode, i.e. a control node and a slave node as shown in Figure 16.3. It is assumed that the slave node is attached to the control node and the control node is always alive to receive the signals from the slave node. As shown in Figure 16.3, the mobile terminal declares the sleep mode and sends sleep mode notification message to let the control node know it enters into the sleep mode. During the sleep mode, the mobile terminal is active only during the short time slot assigned by the control node, and receives the message sent from the control node. The time slot assigned by the control node is called "paging channel". Except the paging channel, the mobile terminal is asleep, and never receives the message from the control node. If the mobile terminal is paged by the base station (control node), the mobile terminal moves to the active mode from the sleep mode and starts the session, e.g. telephone call. Thanks to this mechanism, the mobile terminal does not need to be active all the time, and save lots of energy consumption, thus can lengthen the battery life time.

16.4.1.2 Battery Power Saving by Sleep Mode

The mobile terminal is in the sleep mode when it is waiting to be paged, and it must be woken up by the control node before the mobile terminal starts the communications with the control node. In cellular networks, the paging channel is used for the control node to wake up and page the mobile terminal. This mechanism is illustrated in Figure 16.4, where the receive circuit is active only during the period of the paging channel time slot assigned by the control node, i.e. base station. During the paging channel, RX circuits such as LNA, down converter, demodulator, A/D converters and related baseband circuits are active and the battery power is consumed. They are asleep to save the battery power except the paging channel time slot assigned to the mobile terminal. To implement such sleep mode mechanism, the paging channel needs to be defined and the frame timing synchronization between the slave node and the control node is indispensable to obtain the receiving timing of the paging channel. Note that the energy saving is in proportion to the ratio of T_p/T_f, i.e. the shorter period of T_p compared to the frame period of T_f, the longer the battery life time. To make T_p/T_f larger and to accommodate a large number of mobile terminals in a short paging channel, multi-frame structure is often employed. However, if T_f becomes longer, the connection set-up delay becomes longer. It is a trade-off issue between the connection set-up delay and the battery life time.

On the other hand, regarding the energy saving on the transmitting side of the mobile terminal, it transmits the radio signals if it has data to transmit. If the mobile terminal does not have data to transmit, the mobile terminal may suspend the transmit circuits. During this suspended period, the mobile terminal can cut off the battery power to be fed to the transmit circuits, e.g. HPA (high power amplifier), up converter, modulator, D/A converter and related baseband circuit, to save the battery energy. If CMOS technologies are used in the baseband circuit, it is easy to save energy consumption by cutting off the clock signal, since the power consumption of CMOS LSI is in proportion to the operating frequency. This approach is easy to implement since the mobile terminal knows when it should cut off the battery power or clock signals.

16.4.2 Power Saving in IEEE802.16e Standard

To support battery-operated mobile terminals, IEEE802.16e, i.e. mobile WiMAX supports a power saving mechanism that allows mobile stations to operate for longer durations without recharging the battery [12, 13].

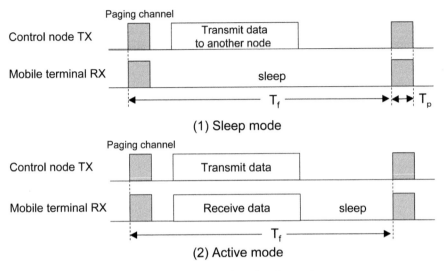

Figure 16.4 Sleep mode and active mode.

As described before, power saving is achieved by turning off some parts of the mobile terminal in a controlled manner when it is not transmitting/receiving data to/from the base station. In IEEE802.16e standard, two power saving modes are defined, i.e. a sleep mode and an idle mode when the mobile terminal is inactive and supports signalling to control the sleep mode and the idle mode of the mobile terminal.

In the sleep mode, the mobile terminal turns itself off and receives the signals from the base station only for the predetermined duration. The periods of absence are negotiated with the base stations. The longer the period, the more the power saving. IEEE802.16e standard defines three power saving classes, i.e. Power Save Class 1, 2, and 3. In Power Save Class 1 mode, the sleep period exponentially increases from a minimum value to a maximum value. The Power Save Class 1 is used by the mobile terminal supporting best-effort and non-real-time traffic. The Power Save Class 2 has a fixed-length sleep period, which is equivalent to the sleep mode in other cellular systems. The Power Save Class 3 supports one-time sleep to be used for the mobile terminal supporting multicast traffic or management traffic.

The other power saving mode, i.e. the idle mode is defined to achieve more power savings. The idle mode allows the mobile terminal to completely turn off and to not be registered with any base stations, but to receive the downlink broadcast traffic. When the downlink traffic arrives for the mobile terminal in the idle mode, the mobile terminal is paged by a set of base stations comprising a paging group. The base station assigns a paging group to the mobile terminal before the mobile terminal enters into the idle mode. The mobile terminal periodically wakes up to update its paging group. As the mobile terminal does not need to register or carry out handover, the idle mode can achieve more power saving than the sleep mode.

16.4.3 Power Saving in IEEE802.11 Standard

Another example of sleep mode in wireless networks is PSM (Power Saving Mode) in IEEE802.11 standard, i.e. wireless LAN (Local Area Networks) [14, 15]. IEEE802.11 is an Ethernet based wireless packet network, which consists of one AP (Access Point) and multiple STAs (Station). AP is a base station to connect multiple STAs to wired infrastructure, where STA is often used as a mobile terminal in conjunction with battery operated PC. Therefore, though paging mechanism as used in cellular networks is not necessary in Ethernet, the sleeping mode needs to be supported in wireless LAN to lengthen the battery life time of STA with PC.

PSM in IEEE802.11 standard is supported in an infrastructure mode of wireless LAN. AP controls power saving of STA using TIM (Traffic Indication Map) in the beacon signal, which is periodically sent to STA as an indicator of which sleeping STA needs to be awaked to receive the packets queued at the AP. The format of TIM element is shown in Figure 16.5. STA wakes up according to the received DTIM (Delivery TIM) and DTIM period if AP has data to transmit to the STA.

In PSM, AP periodically sends beacon signals. In the case of unicast packets, the AP buffers the packets destined to PSM STA from the wired networks and it informs the STA of the buffered information using the PVB (Partial Virtual Bitmap) field in TIM element in the beacon signal. When the STA finds the buffered information of the packets destined to itself in PVB, it starts to receive the packets. PSM STA wakes up to match the cycle of particular beacon signal called DTIM. DTIMs are sent at integral multiples of BI (Beacon Interval), and the integral multiple values of BI, i.e. DTIM Period, is sent as a TIM element in the beacon signal. When the AP send a beacon signal, it decrements the value of the DTIM Period one by one and stores the result in the DTIM Count field. When DTIM Count reaches zero, the beacon signal is sent as DTIM. PSM STA obtains DTIM Period and DTIM Count from the beacon signal and it switches from the sleeping state to active state to receive the packets. In this way, PSM STA wakes up autonomously

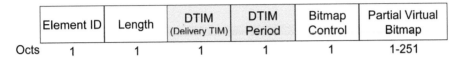

Element ID	Length	DTIM (Delivery TIM)	DTIM Period	Bitmap Control	Partial Virtual Bitmap
Octs 1	1	1	1	1	1-251

Figure 16.5 TIM element format in the beacon signal in IEEE802.11 standard.

at the timing with which DTIM is sent from AP. After receiving the packets, STA switches from the active state to the sleeping state to reduce the power consumption.

16.4.4 Sleep Mode Anywhere

As described in Sections 16.4.1 and 16.4.2, the sleeping mode is indispensable for wireless networks to lengthen the battery life time of the mobile station, and it is widely used in both cellular networks and wireless LAN. The sleep mode is effective to reduce the power consumption if the communications equipment is not always active.

Historically, the sleeping mode has not been used in the wired networks, e.g. conventional Ethernet, because AC power supply can be used to operate non-mobile terminal. However, even in wired networks, the sleeping mode must be effective to reduce the power consumption and to improve the power efficiency in terms of W/bit. Moreover, the sleep mode should be used anywhere, not only in wireless networks but also wired network.

Figure 16.6 illustrates the network architecture and traffic density. Generally speaking, communications networks consist of three kinds of networks. One is the backbone network, where large capacity fiber transmission systems and IP routers are used to carry the transit traffic. Thus, the traffic density is very high in backbone networks. Another network is access network to connect the user to the backbone network. There are various kinds of access networks, e.g. fiber optic based access networks, legacy metallic cable based access networks, fixed wireless access and radio access networks of cellular networks. Then, LAN is connected to the access networks. LAN is a private network owned and operated by the user, and there is a very wide diversity of network scale in LAN from home LAN to enterprise/campus network.

Regarding the network shown in Figure 16.6, the traffic density is relatively high in the backbone network since the traffic from the users are concentrated and carried over the backbone network. On the other hand, the traffic density would become lower in the access network and LAN since each transmission link is dedicated to a user. This means the sleeping mode will be

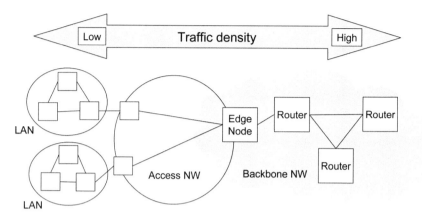

Figure 16.6 Network architecture and traffic density.

effective in LAN and the access networks to achieve high energy efficiency, i.e. low W/bit. Of course, the sleeping mode can be effective in the backbone network since the traffic varies night and day. In this case, it will be effective to introduce the sleeping mode to routers and backbone networks.

As mentioned above, the sleeping mode can be used anywhere to improve the energy efficiency of communications networks. The concept of "Sleep mode anywhere" is one of the most promising approaches for communications networks [11].

16.4.5 Challenges to Sleep Mode Anywhere

As described above, the sleep mode is effective for both wired and wireless networks to reduce the power consumption of network equipment. Though the sleep mode has been mainly used in the mobile stations so far, new challenges have been already initiated in other areas, e.g. Energy Efficiency Ethernet (EEE) [16–18], PSM for AP in wireless LAN [19] and power saving in PON (Passive Optical Network) systems [20]. These new challenges are introduced below.

16.4.5.1 Energy Efficient Ethernet (EEE)

Ethernet link speeds of 100 Mbit/s or 1 Gbit/s are typical nowadays in local area networks. It is obvious that this bandwidth is not used all the time. However, it is reported that Ethernet is used at the full rate less than 5% of the time. This means more than 95% of the energy is wasted. Aiming at

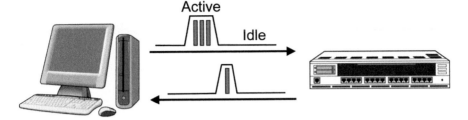

Figure 16.7 Concept of low-power idle.

improved energy efficiency of Ethernet, IEEE802.3 WG started to develop the standard in IEEE P802.3az TG in July 2007.

There are two approaches proposed for Energy Efficient Ethernet. One is Adaptive Link Rate, and the other is Low Power Idle.

In the Adaptive Link Rate scheme, the Ethernet link rate is adapted to match the needs. For example, 100 Mbit/s would be enough for checking e-mail, however, the link rate is changed to 1 Gbit/s for downloading a large file. The power consumption of 100 Mbit/s PHY circuit is less than that of 1 Gbit/s PHY circuit. Therefore, by controlling the Ethernet link rate according to the needs, we can save the energy. The Ethernet interface card works at slower clock rates when the data rate is low and some of the circuits may be turned off. Note that power consumption of CMOS digital circuit is proportional to the clock rate. 10 Gbit/s link uses 10 to 20 W more power than 100 Mb/s links, while 1 Gbit/s link uses about 4 W more. Thus, the power savings will be greater if the links were switching between 10 Gbit/s and 100 Mbit/s.

The problem of the Adaptive Link Rate scheme is the switching time for changing the link rate. When the link rate is changed, the link needs to be re-established. With the current protocol, it would take 2 sec, though the rate switching itself can be performed within a msec. Therefore, the link level protocol needs to be improved to shorten the link establishment time.

The other approach is Low-Power Idle, where the Ethernet card transmits data as fast as possible during the Active state, then returns to Low-Power Idle state. This concept is illustrated in Figure 16.7. We can save energy by cycling between Active and Low Power Idle since power consumption is reduced by turning off unused circuits during Low-Power Idle state. Thus, energy use scales with bandwidth utilization in proportion to the amount of information to be transmitted.

In the Low-Power Idle scheme, Ethernet transmits data at the highest rate and then it moves into a sleep state. During the sleep state, the power for

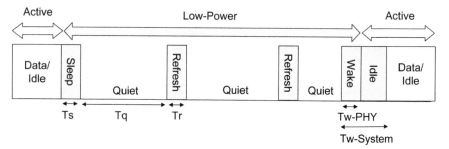

Figure 16.8 Procedure of Low-Power Idle scheme.

MAC layer circuit is turned off to save the energy. One of the challenges of Lowe-Power Idle is to turn on and off an Ethernet card as quickly as possible. For the link rates up to 1 Gbit/s, turning circuits on and off would be easier than switching the link rate from the viewpoint of implementation. Therefore, Low-Power Idle is included in Draft 1.0 of IEEE 802.3az.

Procedure of Low-Power Idle is shown in Figure 16.8, and definitions of terms used are given in Table 16.3. This procedure is based on Master/Slave control. Master enters Low-Power Idle by ceasing transmission and slave acknowledges by also ceasing transmission. After that, master starts refresh cycle by periodically transmitting idle signals. Slave acquires correct timing phase from the received idle signals and responds to master by transmitting idle signals. Then, master ceases transmission and slave acknowledges by also ceasing transmission. In Low-Power Idle state, local/remote PHYs cease transmission and all transmit and receive circuits are turned off to save energy. Local or remote PHY can initiate return to normal operation at any time by transmitting signals, where energy detection is used to detect transmission and turn on all circuits. A refresh cycle and a return to normal operation need to be distinguishable. MAC requests PHY to enter or exit Low-Power Idle. If the remote PHY initiates exit, local PHY immediately signals to the MAC. Master is in control of all timing and initiates entry to Low-Power Idle, refresh cycles, and return to normal operation. Slave can trigger Master to initiate return to normal operation.

16.4.5.2 Power Saving Mode in Wireless LAN Access Point

Although IEEE802.11 standard supports a power saving mode (PSM) for STAs, it does not support that for APs. This is because STAs are often battery-powered, but AP is usually AC-powered. However, from the view point of green communications to save the energy, it is desirable to support PSM, or

Table 16.3 Definitions of terms used in Low-Power Idle.

Term	Description
Sleep Time (Ts)	Duration PHY sends Sleep symbols before going Quiet.
Quiet Duration (Tq)	Duration PHY remains Quiet before it must wake for Refresh period.
Refresh Duration (Tr)	Duration PHY sends Refresh symbols for timing recover and coefficient synchronization.
PHY Wake Time (Tw_PHY)	Duration PHY takes to resume to Active state after decision to Wake.
System Wake TIme (Tw_System)	Wait period where no data is transmitted to give the receiving system time to wake up.

sleep mode for AP. When AP is battery-power operated to be used as a relay station, PSM will be indispensable for AP. Though there are no standardization efforts on PSM for AP, some proprietary proposals have been made aiming at power saving of AP.

One of the proposals on PSM for AP is that AP forces STAs to refrain from transmitting frames by using the network allocation vector (NAV) while AP sleeps. Therefore, AP can be in the sleep state without causing the frame loss at the STAs. Procedure of AP power saving of the proposed scheme is illustrated in Figure 16.9. If the buffer at AP is not empty, AP always stays active. If AP finds the buffer is empty before transmitting DTIM, AP transmits one or multiple CTS-to-self frame and broadcast frame including DTIM. NAV_{max}, NAV_{cts} and NAV_{bro} are defined as maximum NAV period, NAV period in CTS-to-self frame, and the NAV period in broadcasting frame. Let us denote SLEEP the duration of sleep. If SLEEP at AP is longer than NAV_{max}, AP enters into the sleep mode during NAV_{cts}-overhead and continues to transmit CTS-to-self frame till SLEEP becomes less than NAV_{max}. If SLEEP becomes less than NAV_{max}, AP transmits broadcast frame and enters into the sleep mode during NAV_{bro}. During this period, both AP and STA enter into the sleep mode, resulting in power saving of both AP and STA. In this way, the sleep mode is supported for not only STA but also STA.

16.4.5.3 Power Saving in PON Systems

As next generation access network technologies following EPON (Ethernet Passive Optical network) and G-PON (Gigabit PON), 10G-EPON (10 Gigabit EPON) and 10G-PON (10 Gigabit capable PON) have been studied in IEEE802.3av and FSAN/ITU-T. As the power consumption of 10G-EPON and 10G-PON will be much higher than EPON and GPON, power saving is a serious issue. Configuration of PON system is shown in Figure 16.10. Energy

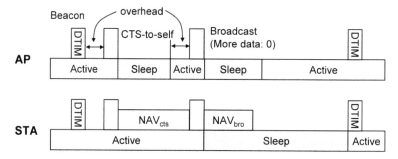

Figure 16.9　Procedure of AP power saving.

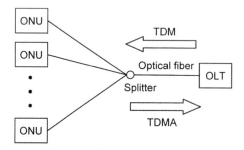

Figure 16.10　Configuration of PON systems.

Efficient Ethernet is to support sleep mode for point-to-point link. However, as PON is a point-to-multi point network using TDM (Time Division Multiplexing) for the downlink and TDMA (Time Division Multiple Access) for the uplink as shown in Figure 1.10, another power saving scheme will be required. One of the examples of sleep mode is that ONU (Optical Network Unit) sleeps periodically if no traffic to OLT (Optical Line Terminal) and ONU asks OLT whether traffic is in the buffer when ONU wakes up. If traffic is in the buffer at OLT, ONU stays active and receives the data. After that, it enters into the sleep mode. This approach is called "Sleep and Periodical Wake-up" or SPW for short.

16.5 Conclusions

In this chapter, it is reviewed how the communications network can be used to improve the energy efficiency of the human life. As seen in the example of the future prediction on the telecommunications traffic in Japan, the traffic of communications networks is rapidly increasing and is expected to grow

340 to 660 times larger in 25 yeas. Although the advanced technologies will help to reduce the power consumption of telecommunications networks, the power consumption of telecommunications network will be more than ten times. Thus, it is strongly required to reduce the power consumption of telecommunications networks equipment, without sacrificing the benefits of telecommunications services since communications network is an essential infrastructure for human life nowadays, not only for traditional voice communications but also for multimedia services.

The green communications is to avoid unnecessary power consumption and to save the energy. In this context, it is pointed out that there are two approaches, i.e. energy saving by communications and energy saving of communications, and we need to consider the trade-off regarding the power saving by communications approach, which is to save the energy required for human life by using communications.

Regarding the power saving of communications, it is stressed that the energy should be used only when they are really needed for communications networks. In this context, the metrics of the energy efficiency of the communications network, i.e. W/bit is proposed and defined. This approach naturally results in the sleep mode as used in cellular networks. The power saving in cellular networks is aiming at long battery life time of the mobile handset. The sleeping mode mechanism of cellular networks and wireless LAN are overviewed to understand how the power saving is performed.

The key of power saving in communications networks is "sleep mode." Based on this observation, the concept of "Sleep Mode Anywhere" is proposed. There are already some activities to save the energy of communications equipment, i.e. Energy Efficient Ethernet, power saving at AP of wireless LAN and power saving of 10G-PON and 10G-EPON. The mechanism of power saving employed them are briefly explained. Note that we should not sacrifice the comfort in dairy life to save the energy by using communications networks. The power consumption of the communication equipment should be reduced keeping the quality of communications services.

References

[1] http://www.ipcc.ch/index.htm.
[2] Climate Change 2007, the IPCC Fourth Assessment Report.
[3] Ken Sakamura, *Ubiquitous Computer Revolution*, Kadokawa Publishing, Tokyo, 2002 (in Japanese).

[4] http://www.ubiq.com/hypertext/weiser/UbiHome.html.

[5] http://www.sics.se/esna/index.php?option=com_frontpage\&Itemid=1.

[6] http://www.soumu.go.jp/menu_02/ict/u-japan/ [in Japanese].

[7] Overview of the smart grid – Policies, initiatives, and needs, ISO New England Inc., February 17, 2009.

[8] IEEE802.15 doc. 15-09-0525-00-004g, Smart utility networks – An international perspective.

[9] http://www.meti.go.jp/press/20071207005/03_G_IT_ini.pdf [in Japanese].

[10] http://itpro.nikkeibp.co.jp/green_it/index.html [in Japanese].

[11] Umehira, M., A perspective on green wireless communications for sustainable society, WPMC2009, Sendai, Japan.

[12] IEEE 802.16e-2005, Part 16: Air interface for fixed broadband wireless access systems – Amendment 2: Medium access control layers for combined fixed and mobile in licensed bands – Corrigendum 1, February 2006.

[13] De Turck, K., Andreev, S., De Vuyst, S., Fiems, D., Wittevrongel, S. and Bruneel, H., Performance of the IEEE 802.16e sleep mode mechanism in the presence of bidirectional traffic. In *Proceedings of 1st International Workshop on Green Communications*, June 2009.

[14] IEEE Std 802.11, Part 11: Wireless LAN Medium Access Control (MAC) and Physical Layer (PHY) Specifications, ISO/IEC 8802-11, 2008.

[15] Qiao, D. and Shin, K.G., Smart power-saving mode for IEEE 802.11 wireless LANs. In *Proceedings of IEEE INFOCOM 2005*, Vol. 3, 13–17 March, pp. 1573–1583, 2005.

[16] IEEE P802.3az Energy Efficient Ethernet Standard Draft D1.4. May 2009.

[17] Reviriego, P., Hernández, J.A., Larrabeiti, D. and Maestro, J.A., Performance evaluation of energy efficient ethernet, *IEEE Communications Letters* **13**(9), September 2009.

[18] Nordman, B., Law, D., Barrass, H., Nels Fuller, J., Sood, K., Olfat, M., Bennett, M., Klein, P., Lanzisera, S. and Diab, W., IEEE 802 tutorial: Energy efficiency and regulation. Available at http://www.ieee802.org/Tutorials.shtml.

[19] Ogawa, M., Hiraguri, T., Nagata, K., Kishida, A. and Umeuchi, M., Performance evaluation of power saving mode in wireless LAN access points, IEICE Technical Report MoMuC2009-12, 2009 [in Japanese].

[20] Kubo, R., Kani, J., Fujimoto, Y., Yoshimoto, N. and Kumosaki, K., Performance analysis of 10 Gigabit class PON systems with power-saving function, IEICE Technical Report CS2009-2, 2009 [in Japanese].

17

Interconnect-Aware High-Level Design Methodologies for Low-Power VLSIs

Michitaka Kameyama and Masanori Hariyama

Department of Computer and Mathematical Sciences, Graduate School of Information Sciences, Tohoku University, 980-8579 Sendai, Japan

Abstract

This paper presents two interconnect-aware high-level optimization techniques. One is scheduling and FU allocation minimizing the total energy time and area constraints based on data-transfer patterns. The other is memory allocation minimizing the number of memory modules and FUs with a parallel access capability for image processing.

Keywords: High-level synthesis, scheduling, memory allocation, functional-unit allocation, interconnection complexity.

17.1 Introduction

In designing state-of-art VLSIs with parallel architecture, one major concern is to reduce complexity of interconnection network between functional units(FUs) and memory modules without degrading the performance.

Firstly, this chapter presents a high-level synthesis approach to minimize the total energy in behavioral synthesis under time and area constraints. The proposed method has two stages, functional unit (FU) energy optimization and interconnect energy optimization. In the first stage, active and inactive energies of the FUs are optimized using a multiple supply and threshold voltage scheme. Genetic algorithm (GA) based simultaneous assignment of supply

R. Prasad et al. (Eds.), Towards Green ICT, 265–274.

and threshold voltages and module selection is proposed. The proposed GA based searching method can be used in large size problems to find a near-optimal solution in a reasonable time. In the second stage, interconnects are simplified by increasing their sharing. This is done by exploiting similar data transfer patterns among FUs. The proposed method is evaluated for several benchmarks under 90 nm CMOS technology. The experimental results show that more than 40% of energy savings can be achieved by our proposed method.

Secondly, this chapter presents an efficient memory allocation to minimize the number of memory modules and FUs with a parallel access capability is presented. An efficient search method is also presented based on regularity of window-type image processing. We give some practical examples including a stereo-matching processor for acquiring 3-dimensional information, and an optical-flow processor for motion estimation. These results demonstrate that the numbers of memory modules are reduced to 2.7 and 10% respectively in comparison with a conventional approach.

17.2 Interconnect-Aware Scheduling and FU Allocation for Energy Minimization [1]

17.2.1 Problem Definition

Our architecture model has FUs with two different supply and threshold voltages. Level converters are used to drive the outputs of low-supply-voltage FUs to high-supply-voltage FUs. The module library has modules with different supply and threshold voltages. The clock gating are used to avoid unnecessary signal transitions. The input for the synthesis flow is a data flow graph (DFG) with N nodes as shown in Figure 17.1(a).

The objective fuction is given by

$$E_{total} = E_{FU}^{active} + E_{FU}^{inactive} + E_{IN} \tag{17.1}$$

where E_{FU}^{active}, $E_{FU}^{inactive}$ and E_{IN} are the active energy of all FUs, the inactive energy of all FUs, and the interconnect energy, respectively. The energy E_{FU}^{active} is given by

$$E_{FU}^{active} = \sum_{i=1}^{N} E_{node\ i}^{active} \tag{17.2}$$

where $E_{node\ i}^{active}$ is the sum of the average active energy per operation of the FU executing the node i and the active energy of the register connected to the

FU. The energy $E_{FU}^{inactive}$ is given by

$$E_{FU}^{inactive} = \sum_{i=1}^{M} \left(E_{FU_i}^{inactive} \times S_{FU_i}^{inactive} \right) \tag{17.3}$$

where $M, E_{FU_i}^{inactive}$ and $S_i^{inactive}$ are the number of FU types, the inactive energy of the FU_i per control step and the total number of control steps where FUs of type FU_i are idle. The energy E_{IN} is originally given by

$$E_{IN} = \sum_{i=1}^{L} \left(\frac{1}{2} C_i \times V_i^2 \times N_i^{trans} \right) \tag{17.4}$$

where L, C_i, V_i and N_i^{trans} are the number of wires, the wire capacitance, the driving voltage of the wire i, and the number of signal transitions at the wire i, respectively. The wire capacitance can be known only after the low-level design tasks such as placement and routing. In order to estimate E_{IN} as accurately as possible in high-level design tasks, *fan-outs* are used since the power dissipation at the wire i increases as the *fan-outs* associated with the wire i increases. Hence, Eq. (17.4) can be rewritten as

$$E_{IN} = \alpha \times \sum FO_i \times V_i^2 \times N_i^{trans} \tag{17.5}$$

where α is the average wire capacitance, and FO_i is the sum of *fan-outs* associated with the wire i. The value α shows how much larger (or smaller) is the interconnect energy compared to the FU energy.

17.2.2 Search Method

For scheduling and FU allocation, the GA-based search method is used to solve the problem in a reasonable time [2]. To consider the interconnect simplification into high-level design flow, binding based on regularity of data transfer is used where common data transfer patterns are exploited and mapped to same interconnects. In order to find the common data transfer patterns, "e-instances" [3] are extracted for a given DFG. An e-instance is a pair of nodes connected by an edge. E-instances are classified into "e-templates" based on the operation types of their source and destination nodes. Figure 17.1(b) shows the e-templates and e-instances derived from the DFG in Figure 17.1(a). E-instances in the same e-template can share the same FUs if their operations are not overlapped. Figures 17.2(a) and 17.2(b) show the

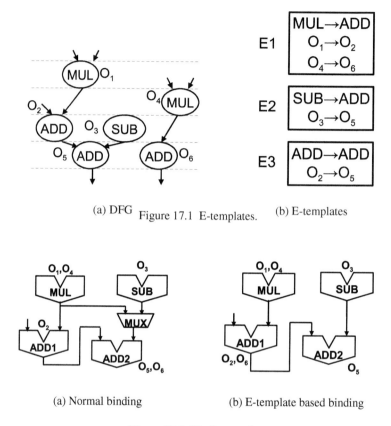

(a) DFG   Figure 17.1 E-templates.   (b) E-templates

(a) Normal binding   (b) E-template based binding

Figure 17.2 Binding results.

binding results. E-template based binding provides a simple interconnection network by sharing the interconnects between O_1 to O_2 and O_4 to O_6. As a result, number of buffers, wires and multiplexers needed for the implementation are decreased and energy consumption in interconnects is reduced. We extend the original concept of E-templates to multiple supply voltages and threshold voltages [1].

Figure 17.3 shows the overall design flow. Since it is difficult to determine the value of α in high level synthesis, the feedback from the layout simulation is used to choose the best solution. The optimization is done for different α values and each layout is evaluated for power consumption using HSPICE.

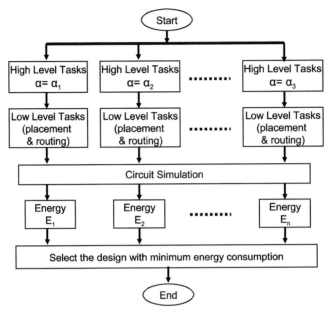

Figure 17.3 Overall design flow.

17.2.3 Evaluation

For the evaluation, 6-metal 1-poly 90nm process with high and low threshold voltage transistors is used. Table 17.1 shows the comparison with two conventional methods: Conv.1 and Conv.2 The method Conv.1 is the the single-supply-voltage scheme with interconnection network simplification [3]. The method Conv.2 is the the dual-supply-voltage scheme without interconnection network simplification. Compared to Conv.1, the proposed method gives 25% average energy savings and 44% maximum energy savings. Compared to Conv.2, up to 36% of energy savings are achieved by interconnection network simplification.

17.3 Memory Allocation for Image Processing on Logic-in-Memory Architecture [4]

17.3.1 Window-Type Image Processing

Our target processing is the window-type image processing where the output/intermediate output depends on a fixed-neighborhood of an input image called "window" as shown in Figure 17.4a. Figure 17.4b shows the DFG of

Table 17.1 Comparisons with conventional methods.

Application	Time constraint	Energy (Conv. 1) [μW]	Energy (Conv. 2) [μW]	Energy (Proposed) [μW]
Auto	1.5T	712.0	654.2	418.9
Regression Filter	2T	525.4	418.9	320.9
FIR Filter	1.5T	712.4	660.9	598.0
	2T	599.6	506.2	489.3
Matrix multiplication	T	4979.3	4337.6	3564.4
	2T	3216.0	2014.0	1803.5

*T = critical path delay

window operation. The labels **A** and **B** on operations denote the operation types of the nodes. This type of processing is frequently encountered in practical image processing such as spatial filter, morphology, and image matching. Moreover, they have the high degree of parallelism. We use window-serial-and-pixel-parallel scheduling where operations are performed in parallel with pixels in a window, whereas operations are performed in a serial manner for windows.

17.3.2 Target Architecture

Figure 17.5 shows the our target architecture called "logic-in-memory architecture" where a single dedicated PE is connected to each memory modules. The advantage of the logic-in-memory architecture is the simple interconnection network between memory modules and PEs. The PEs performs operations of type **A** in parallel, and the outputs are used as inputs of the unit for type-**B** operations.

17.3.3 Optimal Memory Allocation

From a practical point of view, a memory allocation should have a simple address function to map the coordinates of a pixel onto the memory module number. Therefore, we consider a periodical memory allocation where a memory allocation for a whole image is given by repeating a memory allocation for a partial image. Figure 17.6a shows an example of the periodical memory allocation for the window shown in Figure 17.4a. The number on a pixel denotes which memory module the pixel is stored in. This memory

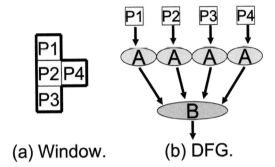

Figure 17.4 Window operation.

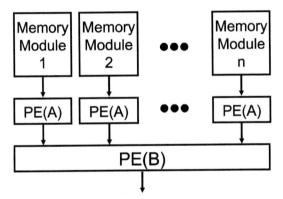

Figure 17.5 Logic-in-memory architecture.

allocation enables parallel access since the pixels in a window are distributed among different memory modules. The periodical memory allocation are represented by two priod vectors U and V. In the periodical memory allocation, the pixels that stored in the same memory modules are given by a linear summation of the period vectors.

Our design objective is to minimize the hardware amount. In the logic-in-memory architecture, the hardware amount is determined by a memory allocation since the number of PEs is determined by the number of memory modules. Therefore, given a window, the optimal memory allocation is defined as the memory allocation that satisfies the following conditions:

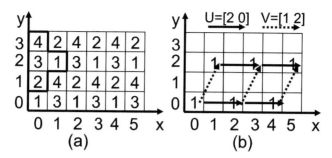

Figure 17.6 Periodical memory allocation.

C1: For an arbitrary location of the window, pixels in the window can be re-trieved in parallel. In other words, the pixels in the window are allocated to different memory modules.

C2: A pixel is allocated to a single memory module to minimize the required memory capacity. This condition ensures that the total memory capacity is minimized.

C3: The number of memory module is minimized. This condition en-sures that the hardware amount is minimized in the logic-in-memory architecture.

C4: The memory allocation is a periodical one. This condition ensures that the hardware for the address function is small.

For example, Figure 17.6a is the optimal memory allocation for the window shown in Figure 17.4.

17.3.4 Search Method

Given the period vectors, the number of memory modules is estimated by the area of the parallelogram made by period vectors $U = [U_x \ U_y]^T$ and $V = [V_x \ V_y]^T$. This is because the memory allocation for a whole image is given by repeating the one for the parallelogram, and the parallelogram must be filled with pixels allocated to different memory modules. The area of the parallelogram is given by

$$S = |U_x \cdot V_y - U_y \cdot V_x|. \tag{17.6}$$

For the example shown in Figure 17.6,

$$S = |1 \cdot 0 - 2 \cdot 2| = 4,$$

Table 17.2 Comparison with the tile-based method.

	Stereo matching		Optical flow extraction	
	Proposed	Tile-based	Proposed	Tile-based
Memory modules	17 (2.7%)	625	10 (10%)	100
PEs	33 (2.6%)	1,249	199 (100%)	199
LUTs	7,951 (7.4%)	106,327	8,840 (9.3%)	94,339
Registers	3,944 (26%)	15,180	487 (200%)	238
Interconnection usage [%]	1 (2.5%)	40	2 (5.4%)	37
Max. Freq. [MHz]	129 (200%)	64	178 (234%)	76
Power[mW]	1,248 (74%,60MHz)	1,679	1,039 (33%,70MHz)	3,119

and S is equal to the number of memory modules. Hence, the memory allocation problem is reduced to finding period vectors with the minimum parallelogram still satisfying the parallel access condition. Once the current minimum number $S_{current}$ of memory modules is obtained in search, the search space for U and V is limited such that $S_{current} > |U_x \cdot V_y - U_y \cdot V_x|$. In this way, the regularity of the periodical memory allocation is fully utilized to reduce the search space.

17.3.5 Evaluation

Let us compare the proposed method with the conventional tile-based memory allocation [5] using two design examples: stereo-matching processor and optical-flow-extraction processor. Table 17.2 summarizes the comparison results. The upper part (memory modules and PEs) and the lower part denote the architectural results and the results of FPGA-based implementations, respectively. StratixIII is used for the FPGA implementations. The search time of the proposed method is less than 1 ms on a PC(Pentium4@2GHz, 1.2G-byte memory) in both designs. As for the stereo-matching processor, the required number of memory modules is reduced to 2.7%, and the required number of PEs is also reduced to 2.7%. This results in great reduction of the interconnection usage in the FPGA implementation. The power dissipation is evaluated at 60MHz since the processor based on the tile-based allocation runs at less than 64 MHz. The power consumption is reduced to 73%. As for the optical-flow-extraction processor, the required number of memory modules is reduced to 10%, and the required number of PEs is same. The power dissipation is evaluated at 75 MHz since the processor based on the tile-based allocation runs at less than 76 MHz. The power consumption is reduced to 33%.

17.4 Summary and Future Directions

This paper describes two interconnect-aware high-level optimization techniques. One is scheduling and FU allocation minimizing the total energy time and area constraints based on data-transfer patterns. The other is memory allocation minimizing the number of memory modules and FUs with a parallel access capability for image processing. The proposed interconnect-aware high-level design methodologies are useful not only for custom VLSIs but also for FPGAs since FPGAs suffers from the large delay and power consumption of the complex switch blocks. Moreover, the proposed memory allocation technique will be a key technology in multimedia processors with highly parallel architectures since most of them adopt highly parallel SIMD architecture and our architecture model is one simplified form of the SIMD architecture.

References

[1] Waidyasooriya, H.M., Hariyama, M. and Kameyama, M., Evaluation of interconnect-complexity-aware low-power VLSI design using multiple supply and threshold voltages, *IEICE Trans. Fund.* **E91-A**(12), 3596–3606, 2008.

[2] Hariyama, M., Aoyama, T. and Kameyama, M., Genetic approach to minimizing energy consumption of VLSI processors using multiple supply voltages, *IEEE Trans. Comp.* **54**(6), 642–650, 2005.

[3] Mehra, R. and Rabaey, J., Exploiting regularity for low-power design. In *Proceedings of the International Conference on Computer-Added Design*, 1996.

[4] Kobayashi, Y., Hariyama, M. and Kameyamaa, M., Optimal periodic memory allocation for image processing with multiple windows, *IEEE Trans. VLSI* **17**(3), 403–416, 2009.

[5] Panda, P.R. and Dutt, N.D., Low-power memory mapping through reducing address bus activity, *IEEE Trans. VLSI* **3**(3), 1999.

18

Development of Wideband Optical Packet Switch Technology

Hideaki Furukawa and Naoya Wada

National Institute of Information and Communications Technology (NICT), Japan

Abstract

Optical packet switch (OPS) systems, which implement high-throughput forwarding of optical packets in the core nodes of "new" generation optical networks, have been investigated. Up to now, we have developed 640 Gbit/s/port interface OPS prototype with multiple optical label processors, high-speed optical switches and optical fiber-delay-line buffers, and demonstrated the error-free operation for 640 Gbit/s wide-colored optical packets by the prototype. This demonstration shows the potential of energy-efficient optical packet switching.

Keywords: Optical packet switch, optical label processeor, optical buffer, wide-colored optical packet.

18.1 Introduction

To cope with the rapid increase of data traffic on networks, it is required to increase the packet throughput in network nodes as well as the transmission capacity in network links. While the link capacity can be increased using various optical multiplexing methods and signal-formats [1, 2], the node-throughput is generally limited due to electronic processing based node-systems. Currently, in electronic routers, wideband optical signals are demultiplexed into lowband electrical signals, and parallel processing is em-

R. Prasad et al. (Eds.), Towards Green ICT, 275–286.

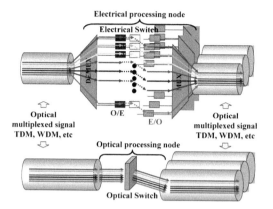

Figure 18.1 Optical processing based node.

ployed. However, such large-scale parallel processing causes serious power consumption problems. Therefore, high-speed and transparent optical processing techniques without optical-to-electrical (O/E) or electrical-to-optical (E/O) converters are quite attractive for node systems as shown in Figure 18.1.

Optical packet switch (OPS) is the key component for high scalability, fine granularity, and more efficient bandwidth utilization "new" generation optical networks [3]. Recently, despite the relative immaturity of optical technologies, many OPS systems have been proposed and demonstrated to implement high-throughput forwarding of optical packets in the core nodes of OPS networks [4–9]. We have also proposed OPS system [4, 8, 9]. One feature of our OPS system is the introduction of optical correlation techniques for label processing. Since the label processing speed is only limited by the propagation delay in the optical device, the throughput of node can be improved. In addition, our OPS system has a possibility to be compatible with wideband optical packets of various formats because the payload's data path, such as a path including switches and buffers, is a purely optical one. Figure 18.2 shows throughput per port of developed OPS prototypes. Rectangular, circle, and triangle marks stand for optical time-division multiplexed (OTDM), wavelength division multiplexing (WDM), and WDM without buffer OPS prototypes, respectively. In 2005, we have developed a 160 Gbit/s/port OTDM based OPS prototype [10, 11] with narrow-band optical label processing, optical switching, high extinction ratio optical buffering, electrical scheduling parts, optical packet transmitter/receiver (Tx./Rx.) [12] and packet bit error ratio tester (BERT) [13]. In order to extend granular-

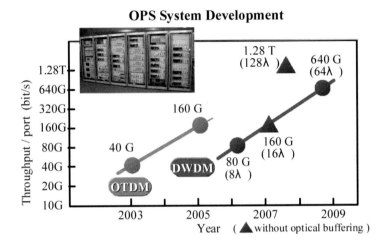

Figure 18.2 Development of optical packet switch by NICT.

ity, scalability, transmission distance, and flexibility of the OPS network, we have introduced a WDM based OPS concept [14]. Up to now, 640 Gbit/s/port WDM based OPS prototype has been demonstrated [15]. In addition, 1.28-Tbit/s/port ultra-wideband OPS without optical buffer has been also demonstrated [16].

18.2 Optical Packet Switch Architecture

Figure 18.3 shows our proposed OPS functional architecture with $N \times N$ input/output ports. The OPS system roughly consists of optical label processing, optical switching, optical buffering, and electrical scheduling. In label processing part, the analysis of a packet label is performed based on parallel optical correlation in the time domain using optical codes (OC), which serve as optical labels. A set of optical correlators works as en/decoders, each storing an OC corresponding to a destination node in a routing table of the OPS network. The encoder generates an OC and the decoder outputs a high-intensity auto-correlation (AC) or low-intensity cross-correlation (XC) signal in the matched or unmatched case between the OC of an optical packet and the decoder's one [9–11]. The payload is spread in the time domain. As a result, the label recognition can be achieved by intensity thresholding. In [9–11], we used en/decoders consisting of some passive devices, such as tunable taps, phase shifters, delays, and combiners on a planar lightwave

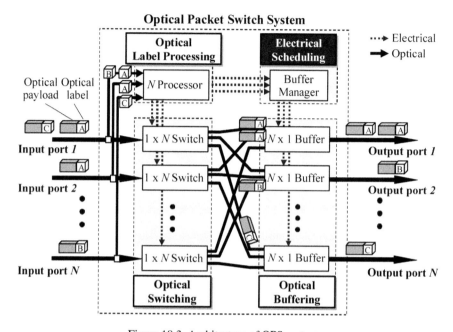

Figure 18.3   Architecture of OPS system.

circuit (PLC). However, since one en/decoder handles only one OC, it is necessary to provide a number of en/decoders in order to handle many labels. This limits the scalability of the OPS system or edge node system.

Here, we introduce multiple optical en/decoders (MOE/MOD) into the OPS system [20]. Although this device has an AWG configuration as shown in Figure 18.4, it is not a wavelength demultiplexer. It behaves like a transversal filter to simultaneously generate and process optical phase-shift keying (PSK) OCs with low latency. To generate a full set of OCs, we send a short laser pulse into one of the device input port, and at the device output ports we obtain sixteen different OCs. Each code is composed of 16 pulses (which are often termed chips in the literature) with a different phase (that is a PSK code). The time interval between two consecutive chips is $\Delta \tau = 5$ ps, so that the code chip-rate is $1/\Delta \tau = 200$ gigachip/s. The label processing speed is $1/(M-1)\Delta \tau = 13.3 \times 10^9$ packet/s. The decoder also outputs an AC or XC signals according to the matched or unmatched case between the OC of an optical packet and the decoder's one at the output ports if an OC is driven into one of the decoder inputs.

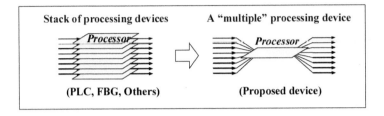

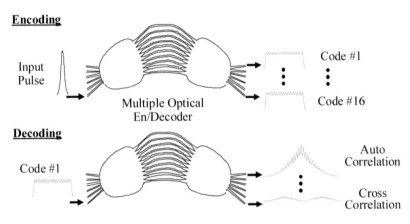

Figure 18.4 Optical label processing by multiple optical en/decoder.

From label processing part, control signals are output to operate optical switches in switching part. Each optical packet is switched toward an appropriate output port according to the label information. For optical switching, we used 1×2 LiNbO$_3$ switches. Since the switching speed is less than 50 ps, even a short-duration optical packet can be efficiency switched with low latency. However, the polarization dependent loss (PDL) is more than 20 dB. The polarization-dependence causes a problem for DWDM-based optical packets. Since each of optical payloads in one packet has different polarization rotation through long-haul fiber transmission, the intensity of all optical payloads is unequal after switching. Moreover, additional devices to adjust the polarization for all payloads lead to complexity of OPS system configuration. Therefore, we adopt recently-developed PLZT optical switches as shown in Figure 18.5a [21], of which polarization independence, proper scale, high-speed (a few ns), low voltage and low power dissipation are promising. The PLZT switch is based on voltage-induced index change with a directional coupler. The PDLs at two output ports are less than 1.2 dB over C-band as shown in Figure 18.5b. This PLZT switch has polarization-independent

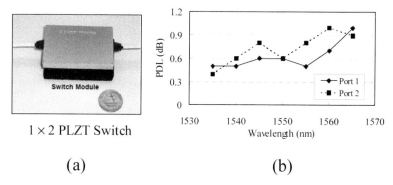

(a)   (b)

Figure 18.5 (a) Photograph of 1×2 PLZT switch module. (b) Polarization dependent loss at output ports 1 and 2 of 1×2 PLZT switch.

electro-optic coefficients, which is obtained by optimizing the composition and the process, and directional couplers with polarization independence enabled by negligible birefringence.

After switching part, there are some possibilities of packet collisions caused by some packets, which are switched toward the same output port at the same timing. To avoid packet collisions, optical buffers delay optical packets during an appropriate time. An optical buffer is made up of an $1 \times M$ optical switch ($1 \times M$ SW) and M optical fiber-delay-lines (FDLs) with different lengths [22]. Figure 18.6 shows the configuration of optical FDL buffers. The length of each FDL is increased by the unit length which can give a delay of one packet duration to optical packets. Each optical packet is switched to a FDL by $1 \times M$ optical switches controlled according to the buffer manager and delayed for appropriate time. In scheduling part, a buffer manager receives arrival information from label processing part, and calculates amount of delay time for each optical packet. Our buffer manager supports synchronous fixed-length packet. Until now, it determined the delays of arriving packets according to first-in-first-out (FIFO) and round robin sequential scheduling [9–11]. In this prototype, to increase the throughput and the number of input/output ports of OPS system, we utilize a buffer manager based on parallel and pipeline algorithm [23] which is an expansion of a simple sequential scheduling. The parallel and pipeline algorithm can provide S times faster processing, where S is the number of input ports of buffer manager, and is also applicable to buffer management of asynchronously arriving variable-length packets.

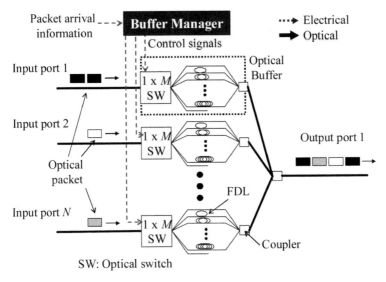

Figure 18.6 Optical FDL buffer configuration.

18.3 Experiment

Figure 18.7 shows the OPS experimental setup and results. EDFAs, band-pass filters and polarization-controller were set in adequate positions. A 10 GHz mode-locked laser diode (MLLD) with a pulse width of 2.0 ps and a center wavelength of 1530.0 nm was used as a label light source. The 10 GHz optical pulse train was modulated according to a packet pattern with a LiNbO3 intensity modulator (LN-IM). The pulses were input into a MOE, and different codes were generated as labels "A" and "B" from two output ports of MOE. The labels "A" and "B" were shifted and combined. Sixty-four arrayed distributed feedback (DFB) lasers with different wavelengths from 1536.7 to 1561.9 nm with 50 GHz spacing were used as light sources of payloads. The continuous waves (CWs) of 32 odd channels were fed into each path of the Tx.1 and Tx.2, and collectively modulated by each LN-IM into 10 Gbit/s NRZ data (PN:2^9-1). The same operation was applied into the CWs of 32 even channels with different data (PN:2^7-1). By coupling labels and 10 Gbit/s payloads of 32 even and 32 odd channels, 640 ($64\lambda \times 10$) Gbit/s packets were generated. The spectrum of generated packets in Tx.1 is shown in Figure 18.7a. The duration of one packet is fixed at 77.1 ns.

Figures 18.7b and 18.7c show the pattern of input packets in an OPS prototype, respectively. We set a MOD to recognize only label "A". In matched

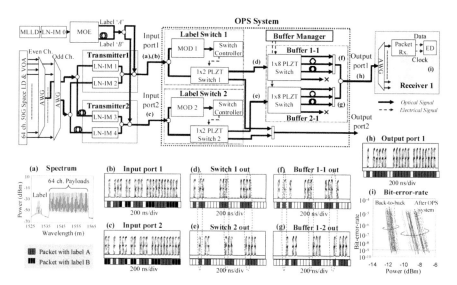

Figure 18.7 Experimental setup and results of 640 Gbit/s WDM-based optical packet switching.

cases of labels "A", a switch controller output gate signals to open and close 1×2 PLZT switches. Figures 18.7d and 18.7e show the switched optical packets with labels "A". The circles show the probability of packet collisions. An optical buffer consists of a 1×8 PLZT switch, 3 FDLs and one drop line. The 1×8 switch was used as 1×4 switch with a gate switch to improve the extinction ratio. The length of 3 FDLs is about 0, 16, and 32 m, respectively. The maximum buffer size is 2 packets. Figures 18.7f, 18.7g and 18.7h show the buffered and output optical packets, respectively. The circles show that packet collisions were avoided. Sixty-four payload-channels were demultiplexed by an AWG. The data and clock of a payload was recovered by an optical packet Rx. and its bit-error-rate (BER) was measured by an error-detector (ED) [13]. The BERs of all 64 payloads were less than 10^{-9}, which was regarded as error free, as shown in Figure 18.7i. The power penalty was 4 dB on average.

We calculated the energy-efficiency-ratio (EER) of our OPS prototypes. The EER is defined as the total number of processable bits per joule [24]. Figure 18.8 shows the EERs versus data-rate per port of each developed OPS prototype with optical buffers. The EERs of 80 Gbit/s/port and 320 Gbit/s/port OPS prototype were obtained in [25, 26], respectively. Since OPS prototype has transparency for various bit-rate by using optical devices, the

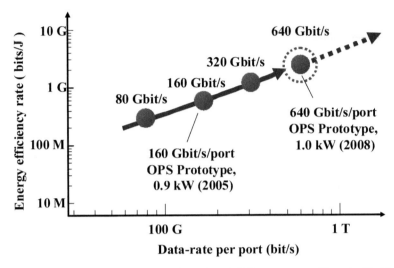

Figure 18.8 Energy-efficiency-ratios of developed OPS prototypes with optical buffers.

data-rate can be quadrupled without the large increase of power consumption (e.g., the power consumption in operation of 640 Gbit/s/port DWDM-based OPS prototype and 160 Gbit/s/port OTDM-based one is 1.0 kW and 0.9 kW, respectively).

18.4 Conclusion

We reported 640 Gbit/s/port OPS prototype. We introduced multiple optical label processors, a parallel pipeline buffer manager, and polarization-independent PLZT optical switches into this prototype. Error-free operation of 640 (64λ × 10) Gbit/s DWDM-based optical packets with the BERs of less than 10^{-9} was achieved by using OPS prototype. These results guarantee that throughput of OPS system can be improved without the large increase of power consumption.

Acknowledgements

The authors would like to thank Professor K. Kitayama of Osaka University, G. Cincotti of University Roma Tre, Dr. K. Nashimoto of EpiPhotonics Corp. for invaluable comments, as well as T. Makino, N, Takezawa, H.

Sumimoto, and Y. Tomiyama of the National Institute of Information and Communications Technology for their support in the experiments.

References

[1] Gnauck, A.H., Charlet, G., Tran, P., Winzer, P.J., Doerr, C.R., Centanni, J.C., Burrows, E.C., Kawanishi, T., Sakamoto, T. and Higuma, K., 25.6-Tb/s WDM transmission of polarization-multiplexed RZ-DQPSK signals, *J. Lightwave Technol.* **26**(1), 79–84, January 2008.

[2] Masuda, H., Sano, A., Kobayashi, T., Yoshida, E., Miyamoto, Y., Hibino, Y., Hagimoto, K., Yamada, T., Furuta, T. and Fukuyama, H., 20.4-Tb/s (204 × 111 Gb/s) transmission over 240 km using bandwidth-maximized hybrid Raman/EDFAs. In *Proc. Optical Fiber Communications Conf. (OFC 2007)*, Anaheim, CA, No. PDP20, 2007.

[3] National Institute of Information and Communications Technology [Online]. AKARI Architecture Design Project, Available: `http://akari-project.nict.go.jp/`.

[4] Kitayama, K. and Wada, N., Photonic IP routing, *IEEE Photon. Technol. Lett.*, **11**(12), 1689–1691, December 1999.

[5] Habara, K., Sanjo, H., Nishizawa, H., Yamada, Y., Hino, S., Ogawa, I. and Suzuki, Y., Large-capacity photonic packet switch prototype using wavelength routing techniques, *IEICE Trans. on Commun.* **E83-B**, 2304–2311, October 2000.

[6] Yoo, S.B., Optical-packet switching and optical-label switching technologies for the next generation optical Internet. In *Proc. Optical Fiber Communications Conf. (OFC 2003)*, Atlanta, GA, Vol. 2, No. FS5, pp. 797–798, 2003.

[7] McGeehan, J., Kumar, S. and Willner, A., Optical time-to-live decrementing and subsequent dropping of an optical packet. In *Proc. Optical Fiber Communications Conf. (OFC 2003)*, Atlanta, GA, Vol. 2, No. FS6, pp. 798-801, 2003.

[8] Wada, N. and Harai, H., Photonic packet routing based on multi-wavelength label switch using multi-section fiber Bragg gratings. *Proc. SPIE*, **4872**, 185–198, 2002.

[9] Wada, N., Harai, H. and Kubota, F., 40 Gbit/s interface, optical code based photonic packet switch prototype. In *Proc. Optical Fiber Communications Conf. (OFC 2003)*, Atlanta, GA, Vol. 2, No. FS7, pp. 801–802, 2003.

[10] Wada, N., Furukawa, H., Fujinuma, K., Wada, T. and Miyazaki, T., 160 Gbit/s/port optical packet switch prototype with variable-length packet BER and loss real-time measurement. In *Proc. 31st European Conference and Exhibition on Optical Communication (ECOC 2005)*, Glasgow, Scotland, Vol. 3, No. We1.4.1, 2005.

[11] Wada, N., Furukawa, H. and Miyazaki, T., Optical code label processing based 160 Gbit/s/port optical packet switch prototype and related technologies, *IEEE J. Select. Topics Quantum Electron.* **13**(5), 1551–1559, September/October 2007.

[12] Wada, N., Iizuka, H., Fujinuma, H. and Miyazaki, T., Prompt-locking optical burst-mode 10 Gbit/s 3R receiver. I *Proc. Asia-Pacific Optical Communication (APOC 2006)*, Gwangju, Korea, No. 6353-10, 2006.

[13] Wada, N., Fujinuma, K., Wada, T., Iiduka, H. and Kubota, T., Pure packet ber and loss real-time measurement with optical label switching and preamble free optical packet 3R. In *Proc. 30th European Conference and Exhibition on Optical Communication (ECOC 2004)*, Stockholm, Sweden, No. PDP-Th4.5.4, 2004.

[14] Harai, H. and Wada, N., More than 10 Gbps photonic packet-switched networks using WDM-based packet compression. In *Proc. 8th OptoElectronics and Communications Conference (OECC 2003)*, Shanghai, China, pp. 703–704, 2003.

[15] Furukawa, H., Wada, N., Harai, H., Takezawa, N., Nashimoto, K. and Miyazaki, T., 640 Gbit/s/port optical packet switch prototype with optical buffer using 1 × 8 PLZT optical switch and parallel pipeline buffer manager. In *Proc. 34th European Conference and Exhibition on Optical Communication (ECOC 2008)*, Brussels, Belgium, no. Tu.3.D.7, 2008.

[16] Wada, N., Kataoka, N., Makino, T., Takezawa, N., Miyazaki, T. and Nashimoto, K., Field demonstration of 1.28T bit/s/port ultra-wide bandwidth colored optical packet switching with polarization independent high-speed switch and all-optical hierarchical label processing. In *Proc. 33rd European Conference and Exhibition on Optical Communication (ECOC2007)*, Berlin, Germany, No. PDP-3.1, 2007.

[17] Awaji, Y., Furukawa, H., Wada, N., Chan, P. and Man, R., Mitigation of transient response of erbium-doped fiber amplifier for traffic of high speed optical packets. In *Proc. Conf. on Lasers and Electro-Optics (CLEO 2007)*, Baltimore, MD, No. JTuA133, 2007.

[18] Furukawa, H., Wada, N., Harai, H., Naruse, M., Otsuki, H., Ikezawa, K., Toyama, A., Itou, N., Shimizu, H., Fujinuma, H., Iiduka, H. and Miyazaki, T., Demonstration of 10 Gbit ethernet/optical-packet converter for IP over optical packet switching network, *IEEE/OSA J. Lightwave Technol.* **27**(13), 2379–2390, July 2009.

[19] Furukawa, H., Wada, N., Awaji, Y., Miyazaki, T., Kong, E., Chan, P., Man, R., Cincotti, G. and Kitayama, K., Field trial of 160 Gbit/s DWDM-based optical packet switching and transmission, *Optics Express* **16**(15), 11487–11495, July 2008.

[20] Wada, N., Cincotti, G., Yoshima, S., Kataoka, N. and Kitayama, K., Characterization of a full encoder/decoder in the AWG configuration for code-based photonic routers – Part II: Experiments and applications, *J. Lightwave Technol.* **24**(1), 113–121, January 2006.

[21] Nashimoto, K., Tanaka, N., LaBuda, M., Ritums, D., Dawley, J., Raj, M., Kudzuma, D. and Tuan, V., High-speed PLZT optical switches for burst and packet switching. In *Proc. of the Second International Conference on Broadband Networks (BROADNETS2005)*, Boston, pp. 195–200, 2005.

[22] Furukawa, H., Harai, H., Wada, N., Miyazaki, T., Takezawa, N. and Nashimoto, K., A 31-FDL buffer based on trees of 1×8 PLZT optical switches. In *Proc. 32nd European Conference and Exhibition on Optical Communication (ECOC 2006)*, Cannes, France, No. Tu4.6.5, 2006.

[23] Harai, H. and Murata, M., High-speed buffer management for 40Gb/s-based photonic packet switches, *IEEE/ACM Trans. Networking* **14**(1), 191–203, February 2006.

[24] Asami, T. and Namiki, S., Energy consumption targets for network systems. In *Proc. 34th European Conference and Exhibition on Optical Communication (ECOC 2008)*, Brussels, Belgium, No. Tu.4.A.3, 2006.

[25] Furukawa, H., Wada, N., Harai, H., Awaji, Y., Naruse, M., Otsuki, H., Miyazaki, T., Ikezawa, K., Toyama, A., Itou, N., Shimizu, H., Fujinuma, H., Iiduka, H., Kong, E., Chan, P., Man, R., Cincotti, G. and Kitayama, K., Demonstration of contention resolution and 100 km transmission for IP/10GbE over 80 Gbit/s (8λ × 10 Gbit/s) colored optical packet switching network using 160 Gbit/s throughput optical packet switch prototype with optical fiber-delay-line-buffer. In *Proc. 33rd European Conference and Exhibition on Optical Communication (ECOC 2007)*, Berlin, Germany, No. 10.5.4, 2007.

[26] Furukawa, H., Wada, N., Harai, H., Takezawa, N. and Miyazaki, T., Demonstration of error-free 320 (32-wavelengths × 10 Gbit/s) Gbit/s wide-colored optical packet switching. In *Proc. 2008 International Conference on Photonics in Switching (PS 2008)*, Sapporo, Japan, No. S-07-3, 2008.

PART III

APPLICATION AND BUSINESS MODELS

19

Application of Green Radio to Maritime Coastal/Lake Communications and Locationing Introducing Intelligent WiMAX (I-WiMAX)

Xiaohua Lian, Homayoun Nikookar and Leo P. Ligthart

International Research Centre for Telecommunications and Radar (IRCTR), Delft University of Technology, 2600 GA Delft, The Netherlands

Abstract

This chapter suggests the introducing of Intelligent WiMAX (I-WiMAX), a green radio technique, able to support new maritime communication services. I-WiMAX utilizes Smart Radio (SR) principles and mobile WiMAX based on the IEEE 802.16e standard. By employing SR, I-WiMAX promises a larger coverage range, higher data rates, more efficient spectrum usage, more reliable communications in sea scenarios, and the prominent capability of localization. I-WiMAX is able to locate ships, and performs highly reliable wireless communications between ships as well as shore. It manages the radio resource in a flexible way, enhances the Quality of Service (QoS) without extra cost of the deployments, consumes less power for communications, and consequently achieves the goals of green radio. Furthermore, it offers the ships and people on shore the versatile broadband services, such as streaming video, wireless Internet access and many others. It is also demonstrated that the green maritime communication system based on I-WiMAX network will significantly improve the efficiency and safety of ship transportation, and sea rescue operation in case of emergency.

R. Prasad et al. (Eds.), Towards Green ICT, 287–307.

Keywords: Green radio, smart radio, Intelligent WiMAX (I-WiMAX), smart antennas, Adaptive OFDM, locationing.

19.1 Introduction

The present broadband technology has fundamentally changed the way to distribute and access information, and consequently it is envisioned to have tremendous market potentials, if it is introduced in a maritime communication environment. This chapter presents the Intelligent WiMAX (I-WiMAX) concept as a new communication architecture, which is dedicatedly designed for green maritime coastal/lake communications and locationing. I-WiMAX is built upon Mobile WiMAX, which is based on the IEEE802.16e standard. The intelligence of I-WiMAX is realized by Smart Radio (SR) and integrates Adaptive Orthogonal Frequency Division Multiplexing (AOFDM) and Smart Antenna (SA) concepts.

With the development of the IEEE 802.16e standard, which is an amendment to 802.16-2004 standards, mobile WiMAX appears to provide high speed data telecommunication services for moving users comparable to the emerging 4G technique. The deployment of Wireless Fidelity (WiFi) on a maritime platform has been presented in [1], where it is demonstrated that in order to meet the requirement of green radio, achieve large range extension and efficient power management, the modification of the Media Access Control (MAC) layer is quite necessary. Furthermore, supplements of the physical layer were also needed, such as a power amplifier for the transmit path. The advantage of WiMAX is that it has a longer distance range and can provide much better performance than WiFi, in terms of coverage, Quality Of Service (QoS) management and spectrum usage efficiency. Thus WiMAX satisfies the basic requirements to become a proper candidate for green radio access networks with high-potential energy efficiency and it can be predicated that WiMAX, unlike WiFi, can offer large communication coverage requirements without extra physical layer supplements.

SR is capable of sensing the communication environment, and consequently can make the radio system adaptive by adjusting its SR parameters. AOFDM adaptively allocates the radio resources, such as the power, the sub-channels and the modulation coding scheme according to Channel State Information (CSI). The flexible modulation scheme of AOFDM, as a developing green technique, will significantly reduce the radiated power requirement. SA recently has been introduced in WiMAX as a "big thing". OFDM is a proper technology and suits well for SA, much more than the existing

3G technology. The IEEE 802.16e standard provides optional features and a signaling structure that enables the usage of SA [2]. A separate point-to-multipoint frames structure can be defined that enables the transmission of downlink and uplink bursts to use directional beams. With specific signal processing techniques allowing for optimizing the adaptive array performance, the Direction Of Arrival (DOA) estimation of ships can be executed accurately. Therefore, with an advanced ranging technique and DOA information, the ability for locationing can be provided as an important service by I-WiMAX usage in maritime communications.

I-WiMAX, as a novel communication system, is aimed to provide a metropolitan access network, and offers higher bandwidth, larger coverage maritime wireless communication networks than the Very High Frequency (VHF) broadcasting system which is currently adopted as Automatic Identification System (AIS) by ships [3, 4]. Since the ships are capable of enjoying a large amount of information anywhere and anytime, the transportation efficiency will be significantly improved by the abundant and versatile information obtained from the new maritime communication system based on broadband wireless networks. Furthermore the locationing ability of the broadband communication system offers position services via Radio Frequency (RF) localization.

This chapter is arranged as follows. Section 19.2 will demonstrate I-WiMAX, as a green radio system for maritime coastal/lake communications and locationing. The following two sections, Sections 19.3 and 19.4, will discuss about the crucial components of SR, AOFDM and SA. The locationing capability of I-WiMAX will be shown in Section 19.5. Section 19.6 gives the examples of the here introduced I-WiMAX application. Finally, Section 19.7 concludes the chapter.

19.2 I-WiMAX

I-WiMAX is built upon SR and mobile WiMAX, as shown in Figure 19.1. SR is a crucial part embedded in I-WiMAX for maritime communication. With SR, the new communication system gains knowledge on the severe sea communication environment, full of echoes and reflections, adapts to it, and therefore increases its cognition and flexibility. The SR technique, which is adopted on top of WiMAX, basically contains two techniques, AOFDM and SA. AOFDM adaptively allocates the radio resource and consequently guarantees the reliable link in the sea channel full of fading and reflection caused by the rough sea surface. SA ensures a higher Signal to Noise Ratio (SNR)

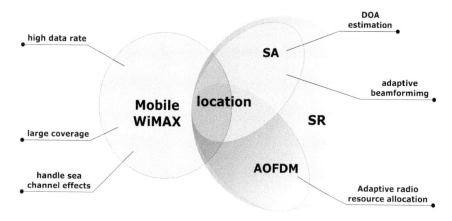

Figure 19.1 I-WiMAX concept for green radio application to maritime coastal/lake communications and locationing.

which consequently results in larger coverage areas by performing adaptive beamforming.

Moreover, with SR, the Based Station (BS) can also achieve locationing abilities. SA for I-WiMAX estimates the DOA of each Subscriber Station (SS) (in our case, most SS are ships) via employing super resolution array signal processing techniques. Together with the Time Of Arrival (TOA) estimation, which measures the distance between the ship and the I-WiMAX BS, the accurate real-time positioning of each moving ship is determined.

In the authors' opinion, it is the first time that I-WiMAX concept is introduced delivering maritime communications services over large areas above water surfaces, although the WiMAX performance in port transportation management was investigated in [5]. WiFi, as a Wireless Local Area Network (WLAN) technique, was presented in [1] to be a Line of Sight (LOS), long range communication solution for marine platforms. Mobile WiMAX, based on IEEE 802.16e, is expected to achieve higher data rates and larger coverage, and is a more promising wireless technology for maritime systems [6]. IEEE 802.16-2004 is very useful in replacing a set of documents all describing different parts of the same technology, with different modification directions. However, after its publication, it still needs an upgrade, mainly for the addition of mobility features. This gave way to 802.16e amendment approved on 7 December 2005 and published in February 2006, which is also known nowadays as mobile WiMAX [7]. The main differences of the IEEE 802.16e standard with regard to IEEE 802.16-2004 standard are the appearances of mobile stations, the Media Access Control (MAC) layer handover procedures, Scalable OFDMA (SOFDMA), Multiple Input Multiple Output (MIMO), data security and others [7]. It is obvious that the IEEE 802.16e standard is dedicated to mobile SS for WiMAX. The coverage and throughput of WiMAX have been subjects of considerable debates, with a throughput of 70 Mb/s and a coverage area of 50 km being claimed as maximum. However, more realistic simulations and trials were run by AT&T in USA and Wireless Broadband (WiBro) in Korea, indicating that the range of mobile WiMAX was about 15 km at a data rate of 20 Mb/s [8].

Another important reason for introducing WiMAX is the adoption of OFDM as its physical layer, which is more suitable for data transmission in sea communications, compared to Single Carrier (SC) Code Division Multiple Access (CDMA), which has been adopted by both Universal Mobile Telecommunication System (UMTS) and High Speed Downlink Packet Access (HSDPA). The use of OFDM increases the data capacity and consequently the bandwidth efficiency with regard to 3G and CDMA. By having carriers very close to each other but still avoiding interferences, due to the orthogonal nature of those carriers, WiMAX presents high spectrum efficiency. In addition, it is also effective against narrow band jamming and interference, as the data are interleaved over multiple carriers. As a result, WiMAX is robust to multipath encountered from the water's reflected surface. However, there are plans to upgrade 3G by introducing OFDM and MIMO in it. The evolution is called, at this moment, Long Term Evolution (LTE).

Locationing ability is a service that can be provided by a maritime wireless communication system for ship navigation and safe transportation. WiMAX adopts OFDM as its physical layer, and consequently it will outperform other systems, which employ narrowband signals, with regard to the resolution of estimation. The super resolution methods for narrow bands will work very well for OFDM signals, taking advantage of the fact that the OFDM signal is able to be divided into several narrowband signals. As a result, DOA estimation for positioning can be performed after the FFT block in the receiver, together with signal decoding, which is less time consuming and needs only a small computational load.

The working procedure of the novel maritime communication approach is illustrated in Figure 19.2. The CSI is fed back to I-WiMAX, and then SR, which belongs to the physical layer of WiMAX, adjusts its operating parameters, including power management, subcarrier selecting, spatial beam directing and so on. SR makes efficient use of all the radio resource according to the changing communication environment, and thus improves the energy

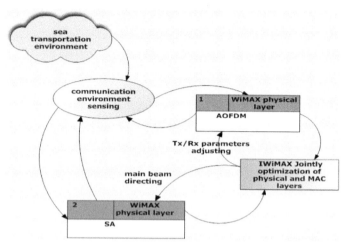

Figure 19.2 Working flow chart of I-WiMAX.

efficiency of green radio. However, in cooperating with SR, WiMAX terminals on ships call for extensions to MAC protocols as described in [1]. For the MAC layer, the required QoS and system throughput should both be satisfied. To maintain high performance of both layers, a MAC-Physics (PHY) cross layer optimized resource allocation scheme is absolutely required. At the end, the optimized operating parameters are fed back to SR.

19.3 AOFDM for I-WiMAX

AOFDM is different from conventional OFDM in the sense that it improves the system performance by adaptive radio resource allocation, where the allocation scheme is based on the understanding of the channel. The resource request and allocation between the BS and each SS is defined in the IEEE 802.16e standard, but the algorithm for allocation the burst to SS is not specified but opened for alternative implementations.

The sea channel is an extremely hash environment for propagating electromagnetic waveforms. RF communications at the sea surface are difficult due to wave blockage, scattering and reflection of RF signals by the surface causing multipath propagation; all result in signal fading and loss. In a severe RF environment, the AOFDM for I-WiMAX automatically reduces the data rate and modulation complexity in order to degrade gracefully instead of ceasing to operate. The modulation scheme can be adjusted based on the

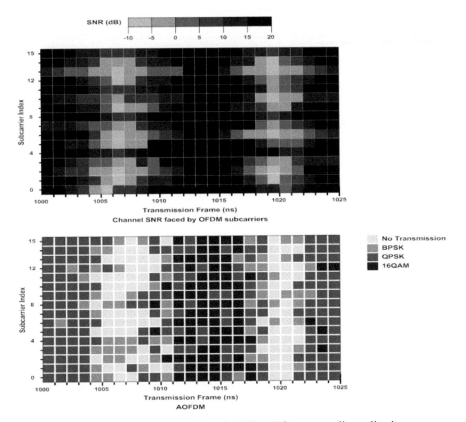

Figure 19.3 Adaptive modulation of I-WiMAX for green radio application.

CSI estimation to guarantee the overall throughput simultaneously supporting good coverage, high data rate and mobility.

Adaptive modulation takes the advantage of the frequency selectivity and time variation by adapting the transmitted signal to match the multipath channel, which is sometimes called "adaptive loading" [10, 11]. AOFDM is able to choose its sub-carriers, which suffer less from fading and face better sub-channels than others. Both the power and data rate in each sub-channel can be adapted. The benefits were described in [9–15]. To achieve the performance advantages of adaptive modulation, however, accurate receiver CSI is required at the transmitter.

An illustration of how AOFDM adjusts data rate and modulation scheme of I-WiMAX for green radio application is shown in Figure 19.3. From this

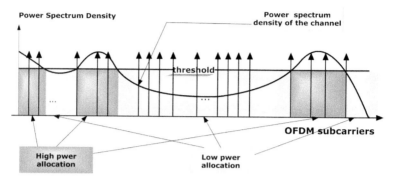

Figure 19.4 Power and subcarriers allocation of I-WiMAX for green radio application.

figure, it can be concluded that the sub-carriers experiencing a good channel with high SNR will be used for higher data rate transmission such as 16 Quadrature Amplitude Modulation (QAM) and Quadrature Phase Shift Keying (QPSK). Meanwhile, to those OFDM sub-carriers which encounter lower SNR, Binary Phase Shift Keying (BPSK) or even non transmission task are assigned to them. An adaptive radio resource allocation method is suggested to be necessary for green communication and therefore is adopted by SR for I-WiMAX, aiming to maintain the estimated Bit Error Rate (BER) and its achievable channel capacity of AOFDM for transmission over a Rayleigh channel [16].

The other radio resources, that can be adaptively allocated, are the power and the selection of sub-carriers. Figure 19.4 illustrates a simple power and subcarriers allocation scheme of I-WiMAX for green radio application. When the Power Spectrum Density (PSD) of the channel is below the defined threshold, those OFDM subcarriers, which locate the actual fading frequency bands, will be assigned to less power. On the other hand, more power will be given to those sub-carriers experiencing good channels. AOFDM assigns the power efficiently to each sub-carrier and saves a large amount of energy for transmitting while maintaining the required QoS. Therefore, it is a most energy efficient scheme for green radio.

Recently several algorithms were proposed for downlink resource allocation in a WiMAX system [17–21] in order to reduce the power for transmission in a green radio concept. An optimum and sub-optimal allocation in terms of maximizing the total downlink throughput was investigated in [17], while the study in [18] proposed a Best Sub-carrier Allocation (BSA) algorithm which uses the feedback of the radio channel quality, and sorts the users to choose subcarriers based on their own channel feedback. More analytical approaches for evaluation of the subcarrier allocation problem are presented in [19,20]. When perfect CSI is not available at the transmitter side, a jointly estimation of the channel and allocation of the resource in OFDMA networks is proposed in [21]. In the case of uplink, based on minimizing the transmitting power, an efficient solution of suboptimal utilization of modulation and coding schemes, defined in an IEEE 802.16 system, is discussed in [22].

19.4 SA for I-WiMAX

SA is an essential technique for green radio because it can, in the transmit mode, focus the energy in the required direction, assist in reducing the multipath reflections and delay spread [23], caused by the fact that a desired signal can arrive via different directions [24]. In the receive mode, it can also perform optimal combining after delay compensation of the incoming multipath signals [25]. Due to the spatial filtering function of SA, co-channel interferences may be reduced at both transmit and receive side [26,27]. When transmitting, the antenna is used to focus the radiated energy in order to form a directing beam in that area, where the receiver is likely to be. The co-channel interference generated in transmitting mode may be further reduced by forming beams exhibiting nulls in the direction of other receivers [28]. The reduction of co-channel interference and the eliminating of multipath fading by using SA may lead to an increased number of users [29,30], which consequently improves the channel capacity and spectrum efficiency [31]. The interference controlling aspect of SA implies energy efficiency for green communication. According to Tsoulos et al. [32], an M-element adaptive array and a multi-beam antenna provide an M-fold increase in antenna gain. This increases the range and reduces the number of base stations required to cover a given area. An adaptive beamformer for the uplink WiMAX compliant system was proposed in [33], which has demonstrated to be able to cope with time varying interference effectively. It was also proved in [1] that with a directional antenna, for example, if a sectorized antenna is employed instead of an omni-directional antenna, the distance tested in a boat-to-air experiment can be at least doubled in WiFi for maritime communications.

Recently an experimental adaptive beamforming system for the IEEE802.16e-2005 OFDMA downlink has been implemented by Motorola labs [34,35]. This system consists of a real time-division duplex link between the BS array antenna and a single antenna SS. The test results show that the

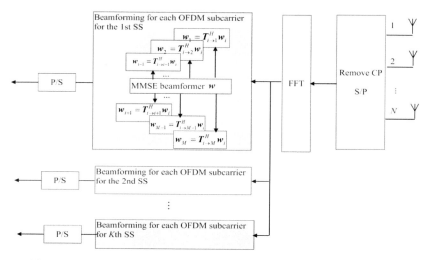

Figure 19.5 Adaptive uplink OFDM beamformer of I-WiMAX for green radio.

Mobile WiMAX is capable of obtaining significant coherent processing gains allowed by array antenna channels. The requirements for supporting SA by WiMAX are discussed in detail in [36], as well as the protocol support and the complexity of extra signal processing as a result of beamforming.

In this section we present an adaptive OFDM beamformer of I-WiMAX for green radio application. Then we show that by employing SA, I-WiMAX will be able to reuse the OFDM sub-carriers, and will therefore significantly improve the spectrum efficiency and also the overall throughput, and consequently achieve the goal of green communication.

19.4.1 Adaptive OFDM Beamformer for Uplink

The adaptive OFDM beamformer of I-WiMAX for green radio is illustrated in Figure 19.5. The Minimum Mean Square Error (MMSE) beamformer is situated after IFFT for transmitting and after FFT for receiving. This is why it is called the post-IFFT/FFT beamformer. Its basic idea is to regard OFDM signals as a combination of several narrowband signals, so that the weights of the MMSE beamformer are decided after the Multicarrier Modulator/Demodulator (IFFT/FFT) for each OFDM subcarrier. The weights for OFDM adaptive beamformer are calculated iteratively. As shown in Figure 19.5, if one set of weight $\mathbf{w}_i$ for the ith OFDM subcarrier is computed, others will be deduced by mapping them into different frequency bands via transformation matrix **T**. Thus the beamforming calculation will be performed only K times for all K SS, which requires less computation load and consumes less time [37].

However, it is not essential to have the MMSE beamformer for the uplink after the IFFT block. Several beamforming approaches exist with varying degree of complexity. The beamforming technique was initially developed in the 1960s for military application with sonar and radar, in order to remove unwanted noise and jamming from the output [31]. The first fully adaptive array was conceived in 1965 by Applebaum in [38], which was designed to maximize the SNR at the array's output. The Least Mean Square (LMS) error algorithm proposed by Widrow in [39] was regarded as an alternative approach to cancelling the unwanted interferences. Further work about LMS techniques was done by Frost in [40] and Griffiths in [41] by introducing the constraints to ensure that the desired signals were not filtered out along with the unwanted signals. However, for stationary signals, both algorithms converge to the optimum Wiener solution [42]. A different technology presented by Capon in 1969 [43] was the Minimum Variance Distortion Response (MVDR) or the Maximum Likelihood Method (MLM). This was the first attempt to automatically localize signal sources. Later in 1974, the Sample Matrix Inversion (SMI) technique, which determines the adaptive antenna array weights directly, was demonstrated in [44]. These all belong to time-dependent reference techniques, which optimize the receive antenna weights in order to identify a known sequence at the output of the antenna array. Another technique is called the spatial reference technique, based on the DOA estimation of signal sources. Many estimation methods of wave number and DOA of signal sources can be found in [45–47]. Blind beamforming is another important beamforming technique, which does not require training sequences and any information concerning the array's geometries, for example, Constant Modulus Adaptive (CMA) beamforming in [48], CM in [49], Spectrum self-COherent REstoral (SCORE) in [41–51] and Decision-Directed Algorithm (DDA) in [52, 53]. To improve the robustness to DOA uncertainty, the Bayesian beamformer was proposed in [54, 55].

With such beamforming technique, the BS as the receiver can detect all signals transmitted by each SS, even if they share the same spectrum. In our case, an efficient spectrum usage of green radio means that all SS are assigned to all OFDM sub-carriers. An adaptive beamforming technique can direct its main beam to the interested signal source, while displaying nulls to the unwanted direction. This ensures that the transmitted signal from one of the SS can be detected and selected by the BS via beamforming. Therefore,

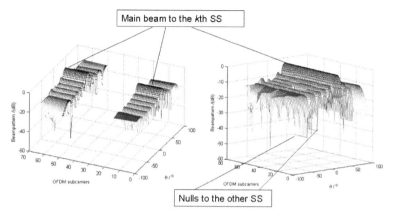

(a) Conventional OFDMA technique   (b) OFDMA technique with beamforming

Figure 19.6 Spectrum reusing via adaptive beamformer of I-WiMAX for green radio.

for those reused OFDM sub-carriers, the weights of them should guarantee the main beam directed towards the correct SS, while at the same time a "null" beam is created towards the other SS. An example can be found in Figure 19.6. The conventional OFDMA technique assigns every SS a set of OFDM sub-carriers as illustrated in figure in Figure 19.6a. Therefore, for each SS, there are several OFDM subcarriers which are not accessible for usage. But with use of the beamforming technique, the SS have the access to those OFDM sub-carriers by putting nulls in the patterns towards users which are also authorized to those sub-carriers; this is shown in Figure 19.6b.

19.4.2 An Adaptive OFDM Beamformer for Downlink

For beamforming in the downlink, the weights for each subcarrier can also be calculated iteratively. Several methods [56–58] have been presented to derive the weights for downlink based on the knowledge of those of uplink, for example, according to the Null Constraint (NC) weights deciding approach. In order to serve more than one user on the single traffic channel, appropriate beamforming must be provided by the BS to satisfy the link quality for all co-channel SS. If we only use the estimated weights for the downlink, the nulls in the patterns towards those co-channel SS are too narrow. Therefore, null broadening methods for beamforming should be introduced [59]. There are several null broadening methods such as high order null broadening, angular spread based approach, and multiple nulling [60–62].

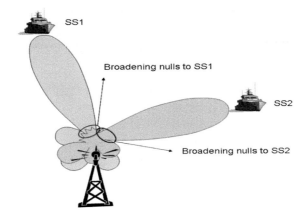

Figure 19.7 Null broadening technique for an adaptive OFDM beamformer for I-WiMAX.

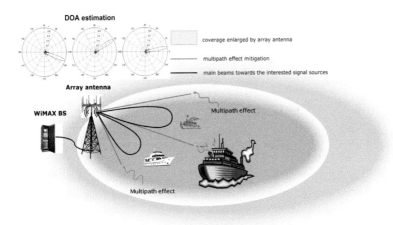

Figure 19.8 DOA estimation for I-WiMAX. (The left top figure shows the SA beam pattern for each DOA estimation.)

With the null broadening technique, the adaptive downlink beamforming of I-WiMAX for green radio is shown in Figure 19.7. Unlike the point nulling technique, the downlink beamforming with null broadening will spread the nulling over an angular region around a certain direction. I-WiMAX performs downlink beamforming with depressed power in a certain angular range with regard to the location of a SS to guarantee the co-channel reusing for the SS.

19.5 Locationing of I-WiMAX

The locationing of I-WiMAX is based on an array signal processing technique. Figure 19.8 illustrates the DOA estimation of the SA for I-WiMAX. It can be seen that the SA main beams are formed in the directions towards the SS while also nulls are created towards the multipath. However, as we discussed before, in green communication in order to achieve the efficiency of the spectrum usage for each user, SA also puts nulls towards other SS.

The array signal processing technique allows for new maritime communications including the detection of ships for deriving and extracting object location information. The ability of source locationing is another important aspect of the array signal processing technique besides SA, referred to as DOA estimation. As discussed before, some Adaptive Array (AA) algorithm depends on the estimated DOA of signal sources. There are various DOA estimation methods [63]. Some of them estimate DOA via the spatial-frequency spectrum, which can be obtained from the Discrete Fourier Transform (DFT) [64] or Maximum Entropy Method (MEM) for spatial sampled signals. The eigen-values of the variance matrix provide a solution to this estimation problem. Multiple Signal Classification (MUSIC) was the first subspace-based method for DOA estimation [65], and was later successfully brought back to the spectrum analysis approach [66, 67]. Some variants of MUSIC include Root-MUSIC [68] and beam-space MUSIC [69]. The Estimation Parameters via the so called Rotation Invariance Technique (ESPRIT) is another subspace-based technique without any requirement of an array manifold vector [70]. The Maximum Likelihood (ML) technique is a well known and frequently used model-based approach in signal processing; as a result, the ML Estimator (MLE) was also introduced to DOA estimation.

These mentioned approaches are applicable for narrow band signal processing. For wide band signal sources, the obvious method is to perform a narrow band decomposition to obtain an array output with a fixed temporal frequency by using DFT or band pass filtering. Examples are the Coherent Signal Subspace Method (CSSM) [71–73], and the Weighted Average of Signal Subspace (WAVES) in [74].

The study in [75] suggests four steps for mobile WiMAX positioning. In [75], the PHY layer performs an important role of detecting the preamble signals and measurement in SS, and selecting different approaches for positioning. The MAC layer is in charge of the signaling and providing opportunities for channel measurement. However, unlike the TOA methods employed by [75], we need an array antenna which is capable of DOA estim-

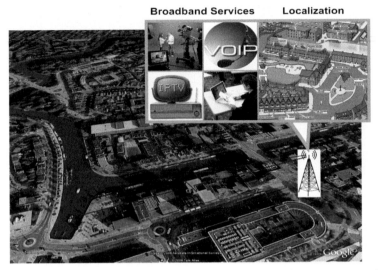

Figure 19.9 Application of I-WiMAX for maritime coastal/lake environment communication (background figure source: Google Earth).

ation. Therefore, the ranging method in [75] with only one BS is considered in our case.

19.6 Application of I-WiMAX

Based on the design of WiMAX, the new maritime communication system is capable of high data rate communication with a larger coverage area. As mentioned in the beginning, the coverage range of WiMAX is expected to be at least 15 km. However, for maritime communication, due to the changeable sea channel conditions, the real coverage may probably shrink somewhat.

The skippers, after introducing the new maritime communication system, will be tended to be broadband users, finding dramatic changes about how to share information, conduct business, and enjoy on board entertainment. The new maritime communication system not only provides faster web surfing and quicker file downloads on boat, but also enables several multimedia applications, such as real-time audio and video streaming, multimedia conferencing, High Definition TV (HDTV), Video on Demand (VoD) and interactive gaming, as shown in Figure 19.9. When ships enter the area covered by the BS on shore, abundant information and new additive services will be offered to skippers and people on board. With the broadband technology, it

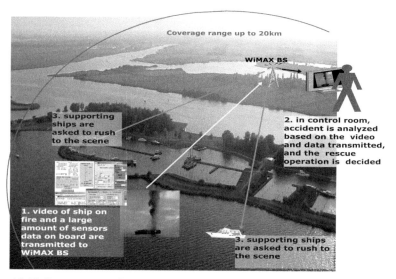

Figure 19.10 Emergency services provided by I-WiMAX (background figure source: Google Earth).

is possible to offer ships guaranteed quality of service as well as specific services types, such as Voice Over IP (VOIP), video images, Internet access, IPTV, and so on. Meanwhile, considering its large coverage, the new maritime communication system needs fewer infrastructures and consequently costs less. The ships are capable of enjoying a large amount of information anywhere and anytime. Furthermore, the locationing ability of the broadband communication system offering position services via RF localization is of great value, as also visualized in Figure 19.9.

Particularly, when an accident happened on the water, the I-WiMAX shows a potential rescue capability as shown in Figure 19.10. Any ship that encounters disasters is required to send the SOS signal, the live video and all the sensors data to the BS. Then on shore in the control room, the experts are able to analyze the cause of the tragedy, decide the best rescuing operation, and monitor the accident scene. At last, several wrecking ships are assigned to the live spot with the determined rescuing operation to help the ship in trouble. The high speed data rate of I-WiMAX ensures the streaming video and a large amount of sensors data to arrive the BS in time, which is valuable in providing emergency services.

19.7 Conclusion

Mobile WiMAX is a fast growing broadband access technology that enables flexible bandwidth and fast link adaptation. In this chapter, we present I-WiMAX based on mobile WiMAX but also employing SR for specifying the adaptation and intelligent radio resources allocation. I-WiMAX can be regarded as a green radio technique that enhances the QoS with less energy consuming and hardware requirement. AOFDM, which is employed by SR, allows for assigning the radio spectrum and power based on the understanding of the wireless channel, as well as choosing the proper modulation scheme. With SA, which is another crucial component of SR, the locationing capability is intended to be provided in I-WiMAX applications. In addition, I-WiMAX is feasible of offering versatile broadband wireless services for ships and will significantly increase the safety of water transportation.

Acknowledgement

The authors acknowledge the support of the Dutch 3TU-Cartesius Institute for this research.

References

[1] Moffatt, C.D., High-data-rate, line-of-sight network radio for mobile maritime communications. In *Proceeding of IEEE Oceans*, pp. 1823–1830, 2005.

[2] Hottinen, A., Kuusela, M., Hugl, K., Zhang, J. and Raghothaman, B., Industrial embrace of smart antennas and MIMO, *IEEE Communication Magazine* 13(4), 8–16, August 2006.

[3] Chang, S.J., Development and analysis of AIS applications as an efficient tool for vessel traffic service. In *OCEANS '04. MTTS/IEEE TECHNO-OCEAN'04*, Vol. 4, pp. 2249–2253, November 2004.

[4] Lessing, P.A., Bernard, L.J., Tetreault, C.B.J. and Chaffin, J.N., Use of the Automatic Identification System (AIS) on autonomous weather buoys for maritime domain awareness applications. In *Proceedings of IEEE Oceans*, pp. 1–6, September 2006.

[5] Joe, J., Hazra, S.K., Toh, S.H., Tan, W. M. and Shankar, J., 5.8G fixed WiMAX performance in a sea port environment. In *Proceedings IEEE Vehicular Technology Conference (VTC)*, pp. 879–883, September 2007.

[6] Nuaymi, L., *WiMAX*, John Wiley & Sons, 2007.

[7] IEEE802.16 working group, http://www.IEEE802.org/16.

[8] WiMAX Telecom, http://www.wimax-telecom.net/en/index.php.

[9] Kalet, I., The multitone channel, *IEEE Trans. Commun.* 37, 119–124, 1989.

[10] Keller, T. and Hanzo, L., Adaptive multicarrier modulation: A convenient framework for time-frequency processing in wireless communications, *Proc. IEEE* **88**, 611–640, May 2000.

[11] Czylwik, A., Adaptive OFDM for wideband radio channels. In *Proc. GLOBECOM*, pp. 713–718, November 1996.

[12] Chung, S.T. and Goldsmith, A.J., Adaptive multicarrier modulation for wireless systems. In *Proc. Asilomar Conf. on Sig., Sys. and Comp.*, pp. 1603–1607, October 2000.

[13] Chow, P., Cioffi, J. and Bingham, J., A practical discrete multitone transceiver loading algorithm for data transmission over spectrally shaped channels, *IEEE Trans. Commun.* **43**(243), 772–775, 1995.

[14] Hanzo, L., Webb, W. and T. Keller, T., *Single- and Multi-carrier Quadrature Amplitude Modulation: Principles and Applications for Personal Communications, WLANs and Broadcasting*, John Wiley & Sons, New York, 2000.

[15] Dardari, D., Ordered subcarrier selection algorithm for OFDM-based high-speed WLANs, *IEEE Trans. Wireless Commun.* **3**(5), 1452–1458, September 2004.

[16] Mallik, R.K., Win, M.Z., Shao, J.W., Alouini, M.S. and Goldsmith, A.J., Channel capacity of adaptive transmission with maximal ratio combining in correlated rayleigh fading, *IEEE Trans. Wireless Commun.* **3**(4), 1124–1133, July 2004.

[17] Zhang, Y.J. and Letaif, K.B., Multiuser adaptive subcarrier-and-bit allocation with adaptive cell selection for OFDM systems, *IEEE Trans. Wireless Commun.* **3**(5), 1566–1575, 2004.

[18] Damji, N. and Le-Ngoc, T., Dynamic downlink OFDM resource allocation with interference mitigation and macro diversity for multimedia services in wireless cellular systems. In *Proceedings of IEEE Wireless Communications and Networking Conference, (WCNC)*, Vol. 3, pp. 1298–1304, March 2005.

[19] Damji, N. and Le-Ngoc, T., Dynamic downlink OFDM resource allocation for broadband multimedia services in wireless cellular systems. In *Proceedings of Broadband Networks, BroadNets*, pp. 589–598, 2004.

[20] Jayaparvanthy, R., Anand, S. and Srikanth, S., Performance analysis of dynamic packet assignment in cellular systems with OFDMA, *IEE Proc.-Communications* **152**(1), 45–52, February 2005.

[21] Curtis, L. and Tuqan, J., An efficient algorithm for channel estimation and resource allocation in OFDMA downlink networks. In *Acoustics, Speech and Signal Processing, IEEE ICASSP*, pp. 3145–3148, March 2008.

[22] Jorguseski, L. and Prasad, R., Downlink resource allocation in beyond 3G OFDMA cellular systems. In *Proceedings of IEEE 18th International Symposium on Personal, Indoor and Mobile Radio Communications, (PIMRC)*, pp. 1–5, September 2007.

[23] Claudio, E.D. and Parisi, R., WAVES: Weighted average of signal subspaces for robust wideband direction finding, *IEEE Trans. Signal Processing* **49**(10), 2179–2190, October 2001.

[24] Barrett, M. and Arnott, R., Adaptive antennas for mobile communications, *IEE Electronic & Communications Engineering Journal* **6**, 203–214, August 1994.

[25] Mizuno, M. and Takeo, O., Application of adaptive array antennas to radio communications, *Electronics and Communications in Japan, Part I* **77**(2), 48–59, March 2007.

[26] Ogawa, Y. and Ohgane, T., Adaptive antennas for future mobile radio, *IEICE Trans. Fundamentals* **E79-A**(7), 961–967, July 1996.

[27] Swales, S.C., Beach, M.A., Edwards, D.J. and McGeehan, J.P., The performance enhancement of multibeam adaptive base-station antennas for cellular land mobile radio systems, *IEEE Trans. on Vehicle Technology* **39**(1), 56–67, February 1990.

[28] Aderson, N. and Howard, P., Technology and transceiver architecture considerations for adaptive antenna systems. In *Proceedings of ACTS Summit*, pp. 965–970, 1997.

[29] Strandell, J., Wennstrom, M., Rydberg, A., Oberg, T., Gladh, O., Rexberg, L., Sandberg, E., Andersson, B.V. and Appelgren, M., Experimental evaluation of an adaptive antenna for TDMA mobile telephony system. In *Proceedings of Personal Indoor and Mobile Radio Communication of (PIMRC)*, pp. 79–84, September 1997.

[30] Monot, J.J., Thibault, J., Chevalier, P., Pipon, F., Mayrargue, S. and Levy, A., A fully programmable prototype for the experimentation of the SDMA concept and use of smart antennas for UMTS and GSM/DCS 1800 network. In *Proceedings of Personal Indoor and Mobile Radio Communication of (PIMRC)*, pp. 534–538, September 1997.

[31] Blogh, J.S. and Hanzo, L., *Third-Generation Systems and Intelligent Wireless Networking smart antennas and adaptive modulation*, Wiley, New York, 2002.

[32] Tsoulos, G.V., Beach, M.A. and Swales, S.C., On the sensitivity of the capacity enhancement of a TDMA system with adaptive multibeam antennas. In *Proceedings of IEEE 47th Vehicle Technology Conference (VTC Spring)*, pp. 165–169, May 1997.

[33] Nicoli, M., Sala, M., Simeone, O., Sampietro, L. and Santacesaria, C., Adaptive array processing for time-varying interference mitigation in IEEE802.16 systems. In *Proceedings of Personal Indoor and Mobile Radio Communication of (PIMRC)*, pp. 1–5, September 2007.

[34] Porter, J.W., Kepler, J.F., Krauss, T.P., Vook, F.W., Blankenship, T.K., Desai, V., Schooler, A. and Thomas, J., An experimental adaptive beamforming system for the IEEE802.16e-2005 OFDMA downlink. In *Proceedings of IEEE Radio and Wireless Symposium*, pp. 475–478, January 2007.

[35] Desai, V., Kepler, J.F. and Vook, F.W., Field data showing the downlink adaptive beamforming gains in an experimental IEEE802.16e-2005 OFDMA system. In *Proceedings of IEEE Radio and Wireless Symposium*, pp. 619–622, January 2007.

[36] Li, Y. and Kenyon, D., An examination of the processing complexity of an adaptive antenna system for WiMAX. In *Proceedings of the 2nd IEE/EURASIP Conference on DSPenabledRadio*, September 2005.

[37] Lian, X., Nikookar, H. and Ligthart, L.P., Adaptive OFDM beamformer with constrained weights for cognitive radio. In *Proceedings of IEEE 69th Vehicle Technology Conference (VTC Spring)*, pp. 1–5, April 2009.

[38] Applebaum, S.P., Adaptive arrays, IEEE Transactions on Antennas and Propagation **24**(5), pp. 585–598, September 1976.

[39] Widrow, B., Mantey, P.E., Griffiths, L.J. and Goode, B.B., Adaptive antenna systems, *Proc. IEEE* **55**, 2143–2159, December 1967.

[40] Frost, O.L., An algorithm for linearly constrained adaptive array processing, *Proc. IEEE* **60**, 926–935, August 1972.

[41] Griffiths, L.J., A simple adaptive algorithm for real-time processing in antenna arrays, *Proc. IEEE* **57**, 1696–1704, October 1969.

[42] Gordara, L.C., Application of antenna arrays to mobile communications, Part II: Beamforming and Direction-of-Arrival considerations, *Proc. IEEE* **85**, 1193–1245, August 1969.

[43] Capon, J., High-resolution frequency-wavenumber spectrum analysis, *Proc. IEEE* **57**, 1408–1418, August 1969.

[44] Reed, I.S., Mallett, J.D. and Brennan, L.E., Rapid convergence rate in adaptive arrays, *IEEE Trans. on Aerospace and Electronic Systems* **10**(6), 853–863, November 1974.

[45] Litva, J., *Digital Beamforming in Wireless Communications*, Artech House, Boston/London, 1996.

[46] Godara, L.C., Application of antenna arrays to mobile communications, Part I: Performance improvement, feasibility, and system considerations, *Proc. IEEE* **85**, 1029–1060, July 1997.

[47] Krim, H. and Viberg, M., Two decades of array signal processing research, *IEEE Signal Processing Magazine*, 67–94, July 1996.

[48] Treichler, J.R. and Agee, B., A new approach to multipath correction of constant modulus signals, *IEEE Trans. on Acoustics, Speech and Signal Processing* **31**(2), 459–472, April 1983.

[49] Godard, D.N., Self-recovering equalization and carrier tracking in a two-dimensional data communication system, *IEEE Trans. Communication* **28**(11), 1867–1875, November 1980.

[50] Agee, B.G., Schell, S.V. and Gardner, W.A., Spectral self-coherent restoral: A new approach to build adaptive signal extraction using antenna arrays, *Proc. IEEE* **78**, 753–767, April 1990.

[51] Wu, Q., Wong, K.M. and Ho, R., A fast algorithm for adaptive beamforming of cyclic signals, *IEE Proceedings on Radar, Sonar and Navigation* **141**(6), 312–318, December 1994.

[52] Henriksson, J.A., Decision-directed diversity combiners-Principles and Simulation results, *IEEE Journal on Selected Areas in Communications* **5**(3), 515–523, April 1987.

[53] Rappaport, T.S., *Wirelss Communications: Principles and Practice*, Prentice Hall, Englewood Cliffs, NJ, 1996.

[54] Bell, K.L., Ephraim, Y. and Van Trees, H.L., A Bayesian approach to robust adaptive beamforming, *IEEE Trans. Signal Processing* **48**(2), 386–398, February 2000.

[55] Lam, C.J. and Singer, A.C., Bayesian beamforming for DOA uncertainty: Theory and implementation, *IEEE Trans. Signal Processing* **54**(11), 4435–4445, November 2006.

[56] Filho, D.Z., Panazio, C.M., Cavalcanti, F.R.P. and Romano, J.M.T., On downlink beamforming techniques for TDMA/FDD systems. In *Proc. Symposio Brasileiro de Telecomunicacoes*, Fortaleza, Brazil, 2001.

[57] Liang, Y.C. and Chin, F.P.S., FDD DS-CDMA downlink beamforming by modifying uplink beamforming weights, *Proceedings IEEE VTS-Fall (VTC)*, pp. 170–174, September 2002.

[58] Liang, Y.C., Chin, F.P.S. and Liu, K.J.R., Downlink beamforming for DS-CDMA mobile radio with multimedia services, *IEEE Trans. on Communication* **49**(7), 1288–1298, July 2001.

[59] Hugl, K., Laurila, J. and Bonek, E., Downlink performance of adaptive antenna with null broadening. In *Proceedings IEEE Vehicular Technology Conference (VTC)*, pp. 872–876, May 1999.

[60] Gershman, A.B., Nickel, U. and Bohme, J.F., Adaptive beamforming algorithms with robustness against jammer motion, *IEEE Trans. on Signal Processing* **45**(7), 1878–1885, July 1997.

[61] Riba, J., Goldberg, J. and Vazquez, G., Robust beamforming for interference rejection in mobile communications, *IEEE Transactions on Signal Processing* **45**(1), 271–275, January 1997.

[62] Steyskal, H., Wide-band nulling performance versus number of pattern constraints for an array antenna, *IEEE Transactions on Antennas and Propagation* **31**(1), 159–163, January 1983.

[63] Kohno, R., Yim, C. and Imai, H., Array antenna beamforming based on estimation on arrival angles using DFT on spatial domain. In *Proceedings of Personal Indoor and Mobile Radio Communication of (PIMRC)*, pp. 38–43, September 1991.

[64] Nagatsuka, M., Ishii, N., Kohno, R. and Imai, H., Adaptive array antenna based on spatial spectral estimation using maximum entropy method, *IEICE Trans. on Communications* **77**(5), 624–633, 1994.

[65] Schmidt, R.O., A signal subspace approach to multiple emitter location and spectral estimation. Ph.D. Thesis, Stanford University, Stanford, CA, November 1981.

[66] Stoica, P., and Nehorai, A., Performance study of conditional and unconditional Direction-of-Arrival estimation, *IEEE Trans. on Acoustics, Speech and Signal Processing* **38**(10), 1783–1795, October 1990.

[67] Stoica, P. and Soderstrom, T., Statistical analysis of MUSIC and subspace rotation estimates of sinusoidal frequencies, *IEEE Trans. on Acoustics, Speech and Signal Processing* **39**(8), 1836–1847, August 1991.

[68] Friedlander, B. and Weiss, A.J., Direction finding for wide-band signals using an interpolated array, *IEEE Trans. on Signal Processing* **41**(4), 1618–1634, April 1993.

[69] Gershman, A. and Amin, M.G., Wideband direction-of-arrival estimation of multiple chirp signals using spatial time-frequency distributions, *IEEE Signal Processing Letter* **7**(6), 152–155, June 2000.

[70] Roy, R., Paulraj, A. and Kailath, T., ESPRIT-estimation of signal parameters via rotational invariance techniques, *IEEE Trans. Signal Processing* **37**(7), 984–995, July 1989.

[71] Wang, H. and Kaveh, M., Coherent signal-subspace processing for the detection and estimation of angles of arrival of multiple wide-band sources, *IEEE Trans. on Acoustics, Speech and Signal Processing* **33**(4), 823–831, August 1985.

[72] Lee, T.S., Efficient wideband source localization using beamforming invariance technique, *IEEE Trans. Signal Processing* **42**(6), 1376–1387, June 1994.

[73] Stoica, P. and Moses, R.L., *Introduction to Spectral Analysis*, Prentice Hall, Upper Saddle River, NJ, 1997.

[74] Claudio, E.D. and Parisi, R., WAVES: Weighted average of signal subspaces for robust wideband direction finding, *IEEE Trans. Signal Processing* **49**(10), 2179–2190, October 2001.

[75] Jiao, W., Jiang, P., Liu, R., Wang, W. and Ma, Y., Providing location service for Mobile WiMAX. In *Proceedings IEEE International Conference on Communications (ICC)*, pp. 2685–2689, May 2008.

20

A Futuristic Outlook on Emerging Business Models

Peter Lindgren, Yariv Taran, Kristin Saughaug and Subria Clemensen

Centre for Industrial Production, Aalborg University, 9220 Aalborg, Denmark

Abstract

In today's complex, knowledge- and innovation-driven, economy innovating business models and their architecture is in growing demand. Innovating business models to become network-based is a complex venture, but critical for the survival of many companies. Business model innovation is not widely researched, though.

This chapter examines the business model and innovation literatures in order to determine whether the business model concept is more than a 'buzz word', adding something fundamentally new and important to management theory and practice. It appears the business model concept is useful, indeed, especially if distinguished from but at the same time related with business strategy and business model innovation in an ongoing developmental cycle. Furthermore, rather than discussing when a change can rightfully be called a business model innovation, it is more useful to consider the space organizations have available to innovate their business model.

Keywords: Innovation, business model, network based business model.

20.1 Introduction

Business models are a challenge to innovators and effective business models are a tremendously valuable asset to a company [11]. Most business leaders

R. Prasad et al. (Eds.), Towards Green ICT, 309–333.

however, when asked to explain their company's business model, would not have a ready answer to give. When they do come up with an answer, they would most likely present their organizational structure and networks. But does this represent a holistic business model? Many managers do not really know what their business model represents and how it functions. Do they even have an explicit business model? And assuming they do, do they know how to continually innovate it successfully for the future?

What is it about business models that make them so difficult for leaders to comprehend? What is really known about the rationale of business models? What is the difference between the innovation of a product and the innovation of a business model? And what is an ICT based business model or an business model based on ICT? Can we draw a distinction between a business model and a ICT based business model?

According to Linder and Cantrell [22], executives cannot even articulate their business models. Many leaders talk about business models but 99 percent have no clear framework for describing their own model. They do know what business they are in, they just cannot describe it clearly. And if they are unable to describe it clearly, they cannot share it effectively throughout their organization. According to Chesbrough [11], 'many try to describe the business model but most are only touching the elephant'.

Magretta [26] continues in this line of thought and argues that both the terms 'business model' and 'strategy' are among the most sloppily used terms in business. 'They are often stretched to mean everything – and end up meaning nothing'. Nonetheless, according to her, these two concepts are of enormous practical value to companies and their future competitiveness.

At present there is extensive knowledge about innovation (see, in general, [43, 46]) and how to innovate products [5, 13, 50] in particular, but very little is known about how to innovate business models and when a business model could be defined as 'new'. So when addressing a futuristic outlook on emerging business models and emerging bsiness model based on ICT – we need both a definition on what is a new business models and how this can be innovated for the future. This concerns both a theoretical and a practical perspective. Some short cases will point to some future examples and possibilities of business models and ICT based business models.

The aim of this chapter, therefore, is to open the discussion on business model innovation as a new field of research, as well as in drawing future potential innovation possibilities, and expected outcomes, when considering how to innovate companies' business models.

20.2 The History and Evolution of the Business Model Theory

What is a 'business' and a 'business model' (BM) really? Unfortunately the answer is inconclusive. Different authors will define the concept in dissimilar ways. In order to simplify things however, it would probably be easier to separate the question into two components: firstly by looking into how the concept of business has been defined and later on what is meant by the 'business model' definition and the functions of the business model.

20.2.1 Towards and Understanding of a Company's Business

Abell in his well known 1980 book *Defining the Business*, argued that a business could be defined according to a three dimensional framework:

1. Customer Functions – customer functions rendered by the company.
2. Customer Groups – customer groups served by the company.
3. Customer Technology – customer technology used to produce customer functions and serve customer groups.

Abell argued that a company's core business, the Strategic Business Unit (SBU), is related to what is inside the box (Figure 20.1). This is related to what the company had defined as its core business. Abell argued that it is particularly important to the company to know about and define the company's core business because of the strong relation of the core business to the company's mission, goals and strategy – the strategic perspective.

The SBU box represents therefore the core business of the company. The space external to the box represents the potential for innovation and further business development. Abell argued that a company, in general, should stick to its core business (core business model). By doing so, the company could achieve optimization of business processes and resource utilizations (operational effectiveness).

When innovating out of the core business area, Abell advised the focal company to carefully analyse its own competences and resources. Abell argued that moving into a new strategic business field (the customer function, customer group or customer technology) would cause a demand for other resources and new competences and, thereby, call for innovation into the field of product development, market development or different degrees of other diversifications.

However, Although Abell's model operates with the three dimensions – which indeed covers and prioritizes some of a company's BM components,

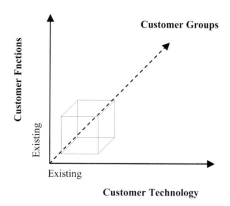

Figure 20.1 Three dimensional framework for the 'business' definition [1].

it explains very little about how the business [model] structural design really looks like, how it is being organized and how it functions.

20.2.2 Definition of the 'Business Model' Concept

The history of clarifying the business model concept is relatively young. However the study has intensified lately. The term 'business model' (BM) has become popular in the mid-1990s during the 'dot com era'. As business ecosystems emerged, many companies have started to rethink their business model and business structure by shifting to an ICT based business model or an 'E-form' business [28].

Many authors, as shown in Table 20.1, have attempted to define the business model concept. Some authors took a narrow (e.g. technological/financial) aspect overview, while others tried to take a more general overview. Some have incorporated strategy as a part of the overall business model, while others left it out. However, it seems that most (if not all) authors agree that a business model is simply a combination of two terms: 'business' and 'model'.

Accordingly, a company's 'business model' serves as a building platform that represents the company's operational and physical manifestation. Thus, the challenge for business model 'designers' is to first identify the key elements and the key relationships that describe the company's 'AS-IS' business model before innovating it.

Table 20.1 Definitions of the term 'business model'.

Timmers, 1998	Business model stands for the architecture for the product, service and information flows, including a description of the various business actors and their roles, the potential benefits for these actors and the sources of revenues … the business model includes competition and stakeholders.
Venkatraman and Henderson, 1998	An architecture along three dimensions: customer interaction, asset configuration and knowledge leverage.
Selz, 1999	A business model is architecture for the firm's product, service and information flows. This includes a description of the various economic agents and their roles. A business model also describes the potential benefits for the various agents and provides a description of the potential revenue flows.
Stewart and Zhao, 2000	Business model is a statement of how a firm will make money and sustain its profit stream over time.
Linder and Cantrell, 2000	The business model is the organization's core logic for creating value.
Hamel, 2000	A business model is simply a business concept that has been put into practice. A business concept has four major components: Core Strategy, Strategic Resources, Customer Interface and Value Network … (Elements of the core strategy include business mission, product/market scope, and basis for differentiation. Strategic resources include core competencies, key assets, and core processes. Customer interface includes fulfillment and support, information and insight, relationships and pricing structure. The value network consists of suppliers, partners and coalitions).
Petrovic et al., 2001	Business model describes the logic of a business system for creating value that lies behind the actual processes.
Weill and Vitale, 2001	A description of the roles and relationships among a firm's consumers, customers, allies and suppliers that identifies major flows of product, information and money and the major benefits to participants.
Magretta, 2002	BMs are stories that explain how the enterprises work…BM describe, as a system, how the pieces of a business fit together, but they don't factor in one critical dimension of performance: competition … a good business model has to satisfy two conditions. It must have a good logic – who the customers are, what they value, and how the company can make money by providing them that value. Second, the business model must generate profits.
Amit and Zott, 2002	A business model is the architectural configuration of the components of transactions designed to exploit business opportunities. The transaction component refers to the specific information, service, or product that is exchanged and/or the parties that engage in the exchange. The architectural configuration explains the linkages among the components of transactions and describes their sequencing.
Osterwalder et al. (2004)	"A blueprint of how a company does business. It is a conceptual tool that contains a set of elements and their relationships and allows expressing a company's logic of earning money. It is a description of the value a company offers to one or several segments of customers and the architecture of the firm and its network of partners for creating, marketing and delivering this value and relationship capital, in order to generate profitable and sustainable revenue stream"
Chesbrough, 2006	The business model is a useful framework to link ideas and technologies to economic outcomes … It also has value in understanding how companies of all sizes can convert technological potential [e.g. products, feasibility, and performance] into economic value [price and profits] … Every company has a business model, whether that model is articulated or not.
Skarzynski and Gibson 2008	The business model is a conceptual framework for identifying how a company creates, delivers, and extracts value. It typically includes a whole set of integrated components, all of which can be looked on as opportunities for innovation and competitive advantage.

20.2.3 Components of the Business Model

As part of the ongoing process of defining the business model concept, many authors have attempted clarifying the components or building blocks as well as constructing a generic BM. Morris et al. [29] presented an overall view

Table 20.2 Nine business model building blocks (Source: Osterwalder et al., 2004 [30]).

Pillar	Business model (the 9 building blocks)	Content and description
Product	Value proposition	Gives an overall view of a company's bundle of products and services.
Customer interface	Target customer	Describes the segments of customers a company wants to offer value to.
	Distribution channel	Describes the various means the company uses to get in touch with its customers.
	Relationship	Explains the kind of links a company establishes between itself and its different customer segments.
Infrastructure management	Value configuration	Describes the arrangement of activities and resources.
	Core competence	Outlines the competencies necessary to execute the company's business model.
	Partner network	Portrays the network of cooperative agreements with other companies necessary to efficiently offer and commercialize value.
Financial aspects	Cost structure	Sums up the monetary consequences of the means employed in the business model.
	Revenue model	Describes the way a company makes money through a variety of revenue flows.

on business models, which was a more interrelated BM framework with a strong focus on an entrepreneurial understanding of business models. They tried to build what they called 'a unified perspective of business models' on top of various academic works that had been carried between the late 1990s and 2003. According to Morris, Schindelhutte and Allen the findings of their cross-theoretical perspective led them to believe that no single theory can fully explain the value creation potential of a business enterprise. Consequently, it is impossible to identify a holistic building block framework for a generic business model depending merely on one author's perspective. However, they did developed an interactive framework that includes three specific decision making levels (*Rules, Proprietary and Foundation*) with relation to six basic 'decision areas' of considerations, namely: Factors related to services and products; Market factors; Internal capability factors; Competitive strategy focus; Economic factors and Growth/Exit factors.

Corresponding with the findings of Morris et al. [29], Osterwalder et al. [30] also developed a framework for business models. Like Morris et al. [29] they too have summed up the academic work of previous business models, and added new theoretical aspects [2–4, 10, 16, 17, 22, 26, 27, 31, 39, 41, 47].

Table 20.2 presents their nine business model 'building blocks'. We used these elements as points of departure for our research on business model innovation.

Additional to Osterwalder et al.'s [30] findings, Osterwalder has further developed those nine building blocks into an ontological business model template as presented in Figure 20.2.

An important question raised but not adequately addressed in the literature is: what is the relation between strategy and business model? Why is

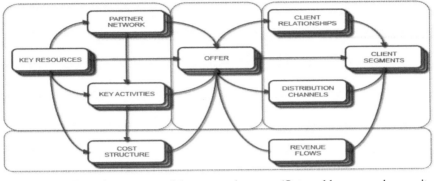

Figure   20.2 Business   model   template   (Osterwalder   web   site: www.privatebankinginnovation.com).

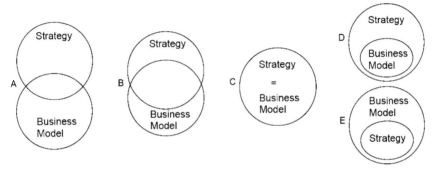

Figure 20.3 Possible overlaps between 'strategy' and 'business model' (Seddon et al., 2004; [37]).

strategy not mentioned in [30]? After all, both Osterwalder et al. and Morris et al. have systematically scrutinized mostly the same literature and yet arrived at different conclusions.

20.2.4 Business Model vs. Strategy

Business model and strategy – both concepts are used frequently by re-searchers as well as company managers, but is it truly possible to define the differences between the two?

Seddon et al. [37] have tackled this issue. Based on the work of vari-ous authors (e.g. [4, 22, 26, 32, 33, 47]), they conclude that a business model should be viewed and defined as an abstract representation (translation) of a company's strategy into a visual blueprint (sketch 'D' in Figure 20.3). This means that, unlike strategy, a business model does not consider the firm's competitive positioning. Furthermore, business models are inward oriented, focusing more on how the firm creates economic value, while strategy is more outward oriented and focuses more on the competitive positioning of the firm.

Osterwalder et al. adopted Seddon et al.'s [37] findings and argued that another difference is that strategy includes execution and implementation, while the business model is more about how a business works as a system. That is why, according to them, some 'winning' business model can be managed badly and fail, just as much as a 'trailing' business model may succeed because of strong management and implementation skills.

Chesbrough and Rosenbloom [10] added two more distinctions. Firstly, business models focus on creating, delivering and capturing value. Strategy is about a company's competitive positioning and sustainability. Secondly, while both are targeted towards creating value, strategy is targeted more towards creating financial value to shareholders, whereas business models are more focused towards creating new value propositions to the business. So, Chesbrough and Rosenbloom seem to conclude that there is a strong overlap between the terms strategy and business model. However they could still present some distinctions between the two concepts (sketch 'B' in Figure 20.3).

20.2.5 When Is a Business Model New?

Having a conceptualization of business models is one thing, innovating them is quite another. According to Magretta [26], new business models are variations on a generic value chain underlying all businesses, which eventually can be divided into two categories:

1. All the activities associated with production; e.g. designing, purchasing and manufacturing.
2. All the activities associated with selling something; e.g. finding and reaching customers, sales transactions and distributing the products/ services.

For that reason, according to her, a new business model can be seen as a new product for unmet needs (new customer segment), or it may focus on a process innovation and a better way of making/selling/distributing an already proven (existing) product or service (to existing and/or new customer segments). Or, formulated more generally, a business model is new if one of the 'building blocks' is new.

According to Amit and Zott [3], business model innovation refers not only to products, production processes, distribution channels, and markets, but also to exchange mechanisms and transaction architectures. Therefore they proposed to complement the value chain perspective by concentrating also on processes that enable transactions. In view of that, they concluded that business model innovation does not merely follow the flow of a product from creation to sale, but also includes the steps that are performed in order to complete transactions. Therefore, the business model as a unit of analysis for innovation potentially has a wider scope than the firm boundaries, since it may encompass the capabilities of multiple firms in multiple industries. Also Chesbrough [11] and the IBM global CEO Study [18] emphasized the importance of business model innovation to appear in the form of organization structure and network relationship changes, such as alliances, joint-ventures, outsourcing, licensing, and spin-offs.

20.2.6 Levels of Business Model Change

The eternal question in the innovation literature is: when can we call something an innovation? Is it new if it is new to the company but not within the industry as such? Is it new if the business model is used elsewhere but not in the local industry? Or should it be completely new to all, like e.g. the Internet and online shopping in the mid 1990s [23]?

The debate on defining incremental or radical innovation (e.g. [21, 35, 43]), concerns the 'how new' question. Incremental innovation involves making small-steps, mainly using existing knowledge to improve, for example, existing products/services, practices and structures. It is short term based and it focuses on the focal company performing better than it already does. Radical innovation, also called quantum leap innovation, is the ability to develop new knowledge and may even be based on competences the company does not possess at the start of the process and result in products/services that cannibalize or remove the basis of existing products/services [43].

Skarzynski and Gibson [38] argued that in order to understand how to innovate a business model, you first need to unpack it into individual components and understand how all the pieces fit together in holistic way. Furthermore, according to them, in order to build a breakthrough business model that rivals will find difficult to imitate, companies will need to integrate a whole series of complementary, value creating components so that the effect will be cumulative.

Another approach to business model innovation, suggested by Linder and Cantrell [22], regards the level of radicality of business model innovation and presents four, what they call, change models: realization models, renewal models, extension models and journey models. In realization models, where most companies are situated, the main issue is to exploit the current potential within an existing operational framework. This business model is considered to be the one with the least actual change like, for example, geographical expansion of the firm, minor changes in a product line, and customer service improvement. Renewal models are firms that leverage their core skills to create a possibly disruptively new position on the price/value curve. Examples include the revitalization of product/service platforms, brands, cost structures and technology bases. Extension models include radical changes by developing new markets, value chain functions, and product/service lines. Finally, journey models involve a complete transformation of the original business model. Here the company moves deliberately and purposefully to a new operating model.

20.2.7 Open Business Model

Chesbrough [11] introduced a whole new way of thinking about business models and innovation of business models in his book *Open Business models – How to Thrive in the New Innovation Landscape*. Firstly, he argued towards open innovation [10] and then opening companies' business models [11]. Until then, all academic work [1, 22, 26, 29, 30] and practically all business thinking had been related to a 'closed' business model – sticking to your core business – where the business model was strongly related to a single, focal company. Chesbrough argued that 'a closed business model' was not efficient for future innovation and a global competitive environment. Much innovation was never even related to the core business model and companies who could have used the results of the innovation were prevented from this because of patents, unwillingness to network and even by lack of knowledge by those who had innovated.

Chesbrough argued that, in the future, companies would have to open up their business models and allow other companies to integrate them with their own business models and even take parts out of the model to use it in other business models. These innovations could be extremely valuable and more effectively used in other business models by other companies.

Because of 'a new global business environment' the business models of companies have to be structured and managed in a more open way – and e.g.

developed and registered patents should become open to other companies of interest in order to support and develop their business models or the innovation of new business models. These innovations could become value adding building blocks to both existing and new business models.

This is the case because companies are playing by different rules using different business models. An open approach will give new opportunities and challenges for both 'giver' and 'taker' of the innovation. It will completely restructure the theory of business models leading to a much more network oriented business model. New business models will focus more on different partners' value equation and will be more open to network partners, both within and outside the value chain.

The open innovation model has become an important concept for further work and development of business models.

20.2.8 ICT as a Key Enabler in New [Virtual] Business Models

ICT, information and communication technology appear to be of ever increasing importance to innovation of new business models and as the backbone of the new business models. First, companies increasingly use ICT to link into new networks. Second, their business models, and especially their new business models, increasingly rely on and are enabled by ICT. Third, their innovation is carried out via advanced ICT tools, which means that companies become increasingly dependent on such ICT networks to extend their business development to the market. Forth more and more of the new business models are based on ICT and operates in ICT based market environment. They are virtual business models based on networks of ICT platforms, software and participants.

Virtual types of business models include, e.g., secondlife.com, google.com, facebook.com and World of Warcraft. Digital types include computers digital networks, network of computers, "cloud of computers", etc. Virtual models include both single company business models and cooperating business models working from different locations [45].

The ICT network based business model of tomorrow includes however business models with many types of ICT structures, including both virtual, digital and physical members.

In a short term perspective, these networks and business models will come under a tremendous pressure of cost, speed, performance and change, especially due to the fact that product and business life cycles in the global market are continuously diminishing [11] but virtual business models even more.

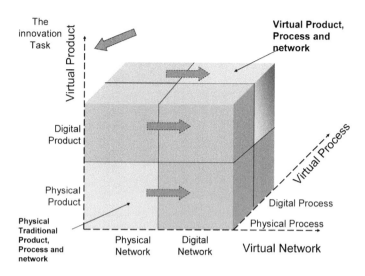

Figure 20.4 From physical to digital to virtual business model based on networks (Lindgren et al. [24] inspired by Whinston et al. [51]).

This has reached a point of no return, where a even strong focus on ICT and virtualization enables companies to cope with this time pressure and the need for more agility, more flexibility and business models which are independent of time, place and people. This is shown as we propose a first generation proposal of virtual business models in Efraim Turbans model showing the distinction between physical, digital and virtual business models.

The original dimensions of electronic commerce [51] is shown in [45, p. 5]. However we expect more advanced virtual business models for the future, which we will comment on a little late in this chapter.

An interesting synergy and spinoff of the above mentioned model is a development from the above mentioned model to a completely new and change in our understanding of the product and service – taking us from the product to the process combinated with the delivery or supplier methods seen as a process.

Business models for the future we expect are not only physical, not only digital, not only virtual but are based on products that are continuously in process, integrated and connected, delivered in a continuously process – where ever and whenever the customer demands it. But not only the product and the delivery change and into a process – touching just two of the building blocks. We predict that all building blocks can and will change in the processes – both individual and across and together the building blocks can change – creating interesting new business models.

This is both the real challenge of the future business models but also be the real potentials of business models of tomorrow.

In a long term perspective continuous improvement, learning and innovation [7], in combination with the continuous need to develop new business models, increases the importance of ICT and strong ICT to make the ability to include and create mega information fast, flexible and dynamically. This will be the platform to a company's and networks ability to innovate new business models, to stay competitive and to keep the business models alive.

So, the answer what is an ICT based business model – well the business models of the future is ICT. We expected all future business models will mainly be based on ICT. However the majority of these future business models will include both physical, digital and virtual business components.

20.3 Discussion

By glancing through the overall research that has been done so far in studying the business model phenomena, a disturbing image is revealed. Clearly, we can conclude that the main reasons to the fuzziness of the 'business model' concept are due to:

1. Inconsistencies regarding business model components.
2. Inconsistencies regarding the relationship between business model and strategy.
3. Inconsistencies regarding the definition of business model innovation and a lack of business model innovation processes (gap).
4. Inconsistencies regarding the ICT's role related to the business model and business model components.
5. Lack of understanding about what is a physical, digital and virtual business model and how to innovate new business models in this context.

Given these circumstances, it can easily be understood why Porter [33] seems to hold such strong opposition to continuing this line of research: the 'business model approach to management becomes an invitation for faulty thinking and self-delusion'. In contrast, Chesbrough, in one of his presentations (Aalborg, 2008), argues that 'we are only started to study the elephant' – a statement that seems to be a 'perfect fit' to our current research findings.

From a theory development perspective, Christensen [12] argues that, if researchers are not able to agree on what the phenomena are, they will have difficulties improving the business model theory on more progressing levels. Accordingly, in order to provide a solid ground for the business model [theory building] research to be built upon, and based on the similarities/differences between the various authors' perspectives, there is a need to narrow down and sharpen the large variation of opinions, as well as to develop new concepts, ideas and categories to the term 'business model' in general, and its innovation processes in particular.

Firstly, as to the components of, what we propose to be, the core of the business model, despite the large variation in opinions we could still identify a strong resemblance between the different components. Thus, based mostly on Osterwalder et al.'s [30] nine building blocks, Amit and Zott's [3] analysis, Chesbrough's [11] open business model innovation, and Hamel [17], we propose the following building blocks (Table 20.3) as the ones that represent best the core components of a business model.

As Table 20.3 suggests, strategy is not included in the core components of the business model. As mentioned earlier, some key researchers, for example Morris et al. [29], Chesbrough [11], Weill and Vitale [47] and Hamel [17], include strategy in their business model definition. It is conceptually clearer, however, also in view of the role of the business model innovation process, to at least analytically distinguish between strategy and the core processes and components of the business model. Strategy, then, involves intentional decision-making on the positioning of the business and the way that strategy will be pursued (outward oriented), while the core business represents the actual implementation of the business strategy (inward oriented).

It is therefore conceptually clearer, as seen in Figure 20.5, to position the business model innovation process between the business strategy (intent) and the core business (realized), and defines business model innovation as the actual process through which the business strategy is realized. Accordingly, as seen in Figure 20.6, it could be argued that strategy, as well as organizational culture and leadership, are embedded within the overall business model template, and thus provides the larger [grounded] platforms for the core business to solidly be based upon.

The next issue concerns the question: when can we call a change in the model a business model innovation? Three approaches have been proposed. The first approach 'defines' business model innovation as a radical change in the way a company does business [11, 18, 19, 22]. Linder and Cantrell in particular are clearly attempting to draw a line in suggesting what can be defined as business model innovation and what is not. The second approach regards any change in any of the [core] building blocks or the relationships

Table 20.3 Core components of the business model (Taran et al., 2009 [42]).

Core question	Core building block
Who do we serve?	**Target customer**/s, market segments and geographies.
What do we provide?	**Value proposition**/s (products and services) that the company offers.
How do we provide it?	[internal] **Value chain** configuration.
	Core competences (assets, processes and activities) that translate company's' inputs into value for customers (outputs).
	Partner network: both strategic partnerships and supply chain management.
	Distribution channel/s that the company uses.
	Customer relationship/s e.g. physical, virtual, etc.
How do we make money?	**Revenue model** – both cost structure and revenue flow.

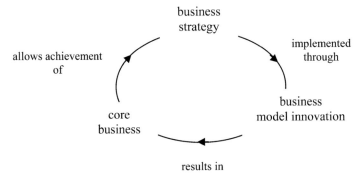

Figure 20.5 Core business, business model innovation and business strategy (Taran et al. 2009, [42]).

between them as a form of business model innovation [3, 26, 30]). A third approach, in line with Abell [1] and Skarzynski and Gibson [38], could be to consider the number of building blocks that are changed simultaneously: any change in one of the building blocks would then constitute an incremental innovation. [Simultaneous] changes in all the building blocks would be the most radical form of business model innovation.

If we combine the second and third approaches with Rogers' (1983) 'scale' of radicality (new to an individual, group, company, industry, society), a three-dimensional space emerges (Figure 20.7), which does not perhaps provide a precise definition of business model innovation, but helps indicate the radicality or, perhaps rather, the complexity of business model innovation. Most importantly, however, is that we get around the discussion on when a change can rightfully be called a business model innovation. Any change is,

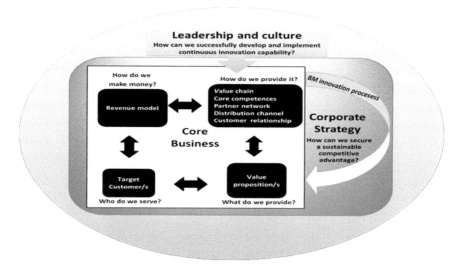

Figure 20.6 Overall business model template (Taran et al., 2009, [42]).

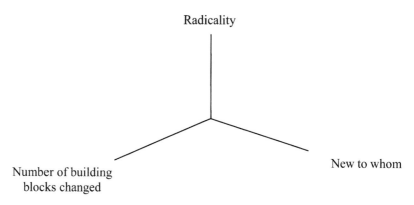

Figure 20.7 A three-dimensional business model innovation scale (Taran et al., 2009 [42]).

but some changes are less complex than others, and some (e.g. radical product innovation, incremental process improvement) are better understood than others (e.g. a holistic, new to the world and a major departure from all business models known so far). This more accurate language for discussing business model innovation will also be of enormous help in classifying any intended business innovation and, based on that, determining and better understanding the managerial issues involved using existing innovation and change theories.

Innovation / Risk mang.	Stage	Stage	Gate 1	Stage	Gate 2	Stage	Stage	Gate 3	Stage	Stage
	AS-IS core business	AS-IS Strategy		Idea generation		Prioritizing	Milestones		Implementation	Dissemination and follow-up
Identify the risks		I		I			I			
Analyse the risks			A		A			A		
Evaluate -prioritise			E		E			E		
Treat the risks				T		T			T	

Figure 20.8 Generic business model innovation process (Taran et al., 2009 [42]).

20.3.1 Building a Process to Business Model Innovation

Our findings so far has illustrated that the business model phenomenon is not well understood while innovation is just partly understood, and the combination of the two leads to a high level of uncertainty and fuzziness, particularly to company managers who will consequently be faced with tremendous challenges and obstacles when attempting to successfully innovate their business model (i.e. financial obstacles, strategic [choices] obstacles, operational obstacles, cultural obstacles, etc.).

Given this, and the understanding that [radical] business model innovation is a 'risky business', is leading to the following tentative theory statement, namely: *Implementation of risk management processes within the overall innovation process is reducing the level of uncertainties, overall costs and [implementation] time, when developing new business models.*

Figure 20.8 represents a generic process that illustrates the possible integration of risk management processes within the overall business model innovation process. The model presented here has adopted Cooper's 'stage-gate' innovation process [13]. However, unlike Cooper, who designed the process to product innovation alone, we related the process to the overall business model innovation process, and designed the model based mostly on our previous findings, presented in Table 20.3, Figures 20.5 and 20.6.

20.3.1.1 Explanation of the Model

The model illustrates a practical implementation process to business model innovation in a linear and systematic manner. The process begins, in stages one and two, with visualizing the core components (Table 20.3) and the 'as-is' strategy of the current operational business model. Then, the process continues by following a 'stage-gate' procedure to implementation and then dissemination of the new business model.

Each gate involves a risk management process, i.e.[1]:

1. **Identify** various risks (i.e. what can be the risk? Under what circumstances can it happen?).
2. **Analyze** the various risks (i.e. what is the likelihood that it will occur?).
3. **Evaluate** those risks (i.e. determine the level of risk that a business is willing to accept).
4. **Treat** the risks (i.e. avoiding/reducing/retaining/transferring … innovating).

The purpose of the 'gates' is therefore to assist with relaxation of constraints, uncertainties and complexities through the business model innovation process itself, as well as in providing more reassurances for company managers regarding the path chosen.

The first risk management 'gate' is drawn, and therefore linked, to the current business strategy. Identifying the risks at this phase will mostly be implied from e.g. a SWOT analysis, where company managers will identify their weaknesses and threats. Then, through careful *analysis* and *evaluation* of each identified [strategic] risk, company managers will search for various possibilities of *treating* those risks, which eventually will result in two possibilities, namely: retrenchment (cost cutting) and/or search for sustained growth via various innovation solutions (e.g. new product/process/service/marketing/position, etc.).

The second risk management 'gate' will begin by identifying various risks to each possible innovation/change, which was proposed at the idea generation phase. Here too, users will follow a systematic process of analyzing, evaluating and then treating those risks, which in this phase will result within the prioritization phase, in choosing which of the ideas proposed will be rejected or approved for further processing.

Finally, the third gate will facilitate identifying, analyzing, evaluating and then treating various risks related to each milestone chosen. The purpose of this gate is therefore to systematically organize the [over] anticipation and sense of urgency, which in many cases can result in a sloppy implementation process that does not take into consideration many of the operational, cultural (readiness), and financial risks involved.

As the model suggests, the processes involved in *identifying* the risks are linked to the previous stage, that is: identify risks with relation to company's

---

[1] There are many risk management processes methods, in this paper we present, as an illustration, a part of NSW risk management model designed by the Department of State and Regional Development, Australia.

Table 20.4 Future trends to BM's innovation.

Context for Innovation	Past trends	Trends for the Future
Market	National Stable Common Mainly physical customer relation	Global Fragmented, Dynamic, Customised New markets (Blue Ocean) More digital and virtual customer relations
Technology	Single technology Expensive Data power low Stable	Mix of technology or multi-technology Cheap Data power over capacity Unstable – Rapid new technology changes
Network	Closed networks, local networks, Fixed networks	Open networks, Dynamic networks, Virtual networks, Global networks
Companies' Competences	Stable competences developed inside the company or in a narrow networks	Dynamic – flexible competences Competences continuously developed under pressure Competences developed with many network partners – sharing core competences and skills in the innovation process (to reduce the risk of disruptive technological changes within the industry)
Products	Mostly physical products To some extent immaterial products Stable product – long life cycle Limited distribution and marketing channels	A mixture of physical, immaterial, digital and virtual products Continuous development of product - short life cycles Many distribution and marketing channels
Product Innovation process	Stable models Slow, linear innovation process	Many product innovation models (flexible models, dynamic models, learning by doing, using, interacting). Rapid prototyping models Lean product innovation process
Success Criteria	Individual success, innovation speed, time to market, cost and performance, local market Emphasis on short term success criteria More emphasis on continuous improvements and managing tangible assets efficiently.	Network-based success, right-speed innovation, time to market, cost and performance, global markets Emphasis on sustainability – short and long term success criteria. More emphasis on radical innovation More emphasis on managing intangible assets efficiently

strategy; [all] relevant ideas proposed, and milestones chosen, respectively; while the processes of *treating* those risks are related to the subsequent phase, namely: treat the strategy weaknesses and threats by proposing possible innovation/change ideas; Treat the large variety of ideas via prioritizing; and treat the milestones risks through the implementation processes, respectively.

20.4 Futuristic Outlook on Emerging Business Models

The business model concept continues to evolve and embrace new perceptions, challenges and opportunities. As part of the authors' preliminary research, important trends and characteristics were found as shown in Table 20.4.

Table 20.4 clearly demonstrates that the rapid changes in the economy are changing within also the rules of the game and companies must in the

Table 20.5 Eight business model innovation possibilities.

Radicality	New to whom	Number of building blocks changed	Expected business model outcome
High	Low	Low	**'Me too' BM.** Follow the others. Second mover strategy. Radical new value, but only to the organization.
High	Low	High	**'Me too' BM.** Despite the radical internal change processes, it is still a second mover strategy.
High	High	Low	Potentially, **Highly competitive BM**. Involved relatively low change management (and risks) processes to be implemented.
High	High	High	**Radical new BM.** Very risky, but also very competitive (blue -ocean)
Low	Low	Low	**Laggard BM.** Not perusing any radical innovation, focus is on continuous improvements, risk of staying behind competition for the long run.
Low	High	High	Potentially, **Highly competitive BM**. Requires high levels of change management (and risks) processes to be implemented.
Low	Low	High	**Laggard BM.** Not perusing any radical innovation, stay behind competition for the long run, and also highly risky due to high levels of incremental innovations and change management processes that are needed to be implemented simultaneously
Low	High	Low	Potentially, **Highly competitive BM**. Involved relatively low change management (and risks) processes to be implemented.

future learn to adjust to those changes – and quickly. The new knowledge based economy is very chaotic and fast moving and the future trajectories are more difficult for the companies to predict. It is a learn based and, therefore, also a customer driven economy. Globalization and competitive advantage are being strongly related to differentiation, and continuous innovation is, in many cases, considered to be implemented via new network formations.

20.5 Business Model Innovation Space

The three dimensions of business model innovation space, presented earlier in Figure 20.7, can be further analyzed into possible expected outcomes, when companies are considering how to innovate their business models, which eventually can result in [roughly] eight possible business model innovation outcomes.

20.6 Future Thinking Is Also about Network-Based Business Model Platforms

Competition today is becoming therefore increasingly difficult for single companies to innovate their business model single handedly. Consequently, a growing tendency of many companies today (and also in the future) is to consider various ways to open their business model toward network-based innovation, particularly when the core knowledge and competences needed to improve their performance is not available in-house [11]. Main potential sources of external competences, among others, involved customers and suppliers, knowledge consultants, venture capitalists and even competitors. Open innovation assist companies to better leverage their internal and external capabilities, as well as in opening new markets and in experimenting with new cutting edge technologies. Consequently, many companies are finding themselves increasingly tied to other firms. The challenge each of them faces is to adjust their own business model to meet their partners' core competences and business models' value, so as to create a new [joint] platform for collaboration and innovation. This tendency, if continued, will have tremendous consequences on the way that companies will compete in the future. Consequently, open business model innovation initiatives are expected to grow significantly in volume (i.e. joint ventures, spin-offs, licensing, outsourcing, and alliances).

Open innovation, which can also be considered as network based innovation platform, may therefore in the future either result from an equal partnership (joint venture), or take place within a loose, more diversified network. In either case, some network partners will be forced to change their individual business logic more radically than others. In addition, the development of a network-based innovation may result in a radical change of customer focus, since it [potentially] involves new technologies, new value propositions, new value chains, new network formations, and new markets.

Furthermore, the new network formation ties will not necessarily be characterized by its industrial homogeneity, but rather by the large diversification of its network partners' identity for the purpose of pursuing radical innovation possibilities.

All in all, the present research and description has led the researchers to believe that the business model concept in the future continues to evolve and embrace new perceptions, challenges and opportunities.

20.7 Conclusion

A lot of buzzwords have come and gone in time, but it seems the business model concept is here to stay. Despite its fuzzy definition and operationalization, it is capturing more and more attention of academics as well as company managers. Consequently, it is being proposed as a new, more sustainable, line of innovation and a new way to meet global competition.

The objective of this chapter was to build a better understanding to business models, ICT based business models and their innovation and, as a result of that, to strengthen researchers' and practitioners' perspectives as to how the process of business model innovation can be realized.

Through the chapter, we introduced and illustrated a framework of future and emerging new business models and the implementation process together with the challenges of these BM innovations. In our research of business models we have enhanced our views and understanding of the BM concept. It is the researchers' belief that the future BM concept, particularly with relation to its innovation, in the future serves more than the translation of a single company's BM and strategy into a visual blueprint. It should consider a managerial tool for new exploration possibilities, mostly enabled by its relation to network, green information and communication, sustainability and the global market and growth.

A new understanding and theory of business models and what they are all about seems to be developing quickly at present. It is the researchers believe that the future BM concept should be considered and analyzed also from a network perspective. A more network based business model moving from a closed to open business models is developing, with much more focus on values, processes, cost – not only of partners inside the business models but also those outside – partners who are either in the first, second or even third levels of the chain of customers. Business models which are not only physical but also digital, virtual or a combination of these.

Accordingly, it is the authors belief, that it is not sufficient to visualize who the network partners of the firm are, as suggested by previous academic work on the single company BM framework, but also to draw a framework of how the network-based BM should be formed as well as in understanding how to [and who should] lead the network BM successfully and how to innovate it? However, more research is needed in order to be able to draw more concrete conclusions regarding the network-based business model phenomena.

References

[1] Abell, D.F., *Defining the Business: The Starting Point of Strategic Planning*, Prentice-Hall, 1980.

[2] Afuah, A. and Tucci, C., *Internet Business Models and Strategies*, Boston: McGraw Hill, 2003.

[3] Amit, R. and Zott, C., Value creation in e-business, *Strategic Management Journal* **22**(6/7), 493–520, 2001.

[4] Applegate, L.M., E-business models: Making sense of the internet business landscape. In G. Dickson, W. Gary and G. DeSanctis (Eds.), *Information Technology and the Future Enterprise: New Models for Managers*, Prentice Hall, Upper Saddle River, NJ, 2001.

[5] Baker, M. and Hart, S., *Product Strategy and Management*, Prentice Hall, Harlow, pp. 157–196, 2007.

[6] Barney, J.B., Firm resources and sustained competitive advantage, *Journal of Management* **17**, 99–120, 1991.

[7] Bohn, K.R. and Lindgren, P., *Right Speed in Network-Based Product Development and Its Relationship to Learning*, Continuous Improvement (CIM) and Continuous Innovation (CI), 2003.

[8] Burt, R.S., *Structural Holes: The Social Structure of Competition*, Harvard University Press, Cambridge, MA, 1992.

[9] Carlile, P.R. and Christensen, C.M., The cycles of theory building in management research, Version 6.0, Working Paper, Innosight, 2005, www.innosight.com/documents/Theory%20Building.pdf.

[10] Chesbrough, H. and Rosenbloom, R.S., The role of the business model in capturing value from innovation: Evidence from XEROX Corporation's technology spinoff companies, *Industrial and Corporate Change* **11**(3), 529–555, 2000.

[11] Chesbrough, H., *Open Business Models. How to Thrive in the New Innovation Landscape*, Harvard Business School Press, 2006.

[12] Christensen, C.M., The ongoing process of building a theory of disruption, *Journal of Product Innovation Management* **23**, 39–55, 2006.

[13] Cooper, R., *Winning at New Products: Accelerating the Process from Idea to Launch*, 2nd edition, Addison-Wesley, Boston, MA, 1993.

[14] Doz, Y. and Hamel, G., *Alliance Advantage*, Harvard Business Press, Boston, MA, 1998.

[15] Dyer, J. and Singh, H., The relational view: Cooperative strategy and sources of inter-organizational competitive advantage, *Academy of Management Review* **23**, 660–679, 1998.

[16] Gordijn, J., Value-based requirements engineering – Exploring innovative e-commerce ideas, PhD Thesis, Vrije Universiteit Amsterdam, the Netherlands, 2002.

[17] Hamel, G., *Leading the Revolution*, Harvard Business School Press, Boston, MA, 2000.

[18] IBM, Expanding the innovation horizon, 2006, http://www-935.ibm.com/services/uk/bcs/html/t_ceo.html.

[19] IBM, The enterprise of the future, 2008, http://www-935.ibm.com/services/us/gbs/bus/html/ceostudy2008.html.

[20] Kuhn, T., *The Structure of Scientific Revolutions*, University of Chicago Press, Chicago, 1962.

[21] Leifer, R., *Critical Factors Predicting Radial Innovation Success*, Technische Universität, Berlin, 2002.

[22] Linder, J. and Cantrell, S., *Changing Business Models: Surveying the Landscape*, Accenture Institute for Strategic Change, Cambridge, 2000.

[23] Lindgren, P., Taran, Y. and Schmidt, A.M., The analytical model for NEWGIBM. In P. Lindgren and E.S. Rasmussen (Eds.) *New Global ICT-Based Business Models*, Aalborg Universitetsforlag, 2008.

[24] Lindgren, P., Taran, Y. and Boer, H., From single firm to network-based business model innovation, *International Journal of Entrepreneurship and Innovation Management*, Proceedings, 2009.

[25] Loukis, E. and Tavlaki, E., Business model: A perquisite for success in the network economy. In *Proceedings of 18th Bled eConference: eIntegration in Action*, Bled, Slovenia, 6–8 June 2005.

[26] Magretta, J., Why business models matter?, *Harvard Business Review* **80**(5), 86–92, 2002.

[27] Mahadevan, B., Business models for internet-based e-commerce. An anatomy, *California Management Review* **42**(4), 55–69, 2000.

[28] Moore, J.F., The new corporate form. In D. Tapscott, A. Lowy, and D. Ticoll (Eds.), *Blueprint To The Digital Economy – Creating Wealth in the Era of E-Business*, McGraw-Hill, New York, 1998.

[29] Morris, M., Schmindehutte, M. and Allen, J., The entrepreneur's business model: Toward a unified perspective, *Journal of Business Research* **58**(6), 726–735, 2003.

[30] Osterwalder, A., Pigneur, Y. and Tucci, L.C., Clarifying business models: Origins, present, and future of the concept, *Communications of AIS* **16**, 1–25, 2004.

[31] Petrovic, O., Kittl, C. and Teksten, R.D., Developing business models for e-business. In *Proceedings of the International Conference on Electronic Commerce*, Vienna, October–November 2001.

[32] Porter, M.E., *Competitive Advantage: Creating and Sustaining Superior Performance*, Free Press, New York, 1985.

[33] Porter, M.E., Strategy and the internet, *Harvard Business Review* **79**(3), 63–78, 2001.

[34] Rogers, E.M., *Diffusion of Innovations*, 3rd edn, The Free Press, New York, 1983.

[35] Rosenau, J.N., Turbulent change. In P.R. Viotti and M.V. Kauppi (Eds.), *International Relations Theory: Realism, Pluralism, Globalism*, Macmillan Publishing Company, New York, pp. 438–448, 1993.

[36] Schumpeter, A.J., *Capitalism, Socialism and Democracy*, Harper & Row, New York, 1942.

[37] Seddon, P., Lewis, G. and Shanks, G., The case for viewing business models as abstractions of strategy, *Communications of the Association for Information Systems* **13**, 427–442, 2004.

[38] Skarzynski, P. and Gibson, R., *Innovation to the Core*, Harvard Business School Publishing, Boston, MA, 2008.

[39] Stahler, P., Geschäftsmodelle in der digitalen Ökonomie. Merkmale, Strategien und Auswirkungen, PhD Thesis, University of St. Gallen, Switzerland, 2001.

[40] Stewart, D.W. and Zhao, Q., Internet marketing, business models, and public policy, *Journal of Public Policy & Marketing* **19**, 287–296, 2000.

[41] Tapscott, D., Ticoll, D. and Lowy, A., *Digital Capital. Harnessing the Power of Business Webs*, Harvard Business School Press, Boston, MA, 2000.

[42] Taran, Y., Boer, H. and Lindgren, P., Theory building – Towards an understanding of business model innovation processes. In *Proceedings of the International DRUID-DIME Academy Winter Conference, Economics and Management of Innovation, Technology and Organizational Change*, 2009.

[43] Tidd, J., Bessant, J. and Pavitt, K., *Managing Innovation. Integrating Technological, Market and Organizational Change*, John Wiley & Sons, Chicester, 2005.

[44] Timmers, P., Business models for electronic markets, *Journal on Electronic Markets* **8**(2), 3–8, 1998.

[45] Tuban, E., *Electronic Commerce – A Managerial Perspective*, Pearson, Chapter 1, p. 5, 2008.

[46] Ulrich, K.T. and Eppinger, S.D., *Product Design and Development*, McGraw-Hill, New York, 2000.

[47] Weill, P. and Vitale, M.R., *Place to Space*, Harvard Business School Press, Boston, MA, 2001.

[48] Wernerfelt, B., A resource-based view of the firm, *Strategic Management Journal* **5**, 171–180, 1984.

[49] Williamson, O.E., The economics of organization: The transaction cost approach, *Am. J. Sociol.* **87**(4), 77–548, 1981.

[50] Wind, Y., A new procedure for concept evaluation, *Journal of Marketing* **37**, 2–11, 1973.

[51] Whinston, A.B. Stahl, D.O. and Choi S., *The Economics of Electronic Commerce*, Macmillan Technical Publishing, Indianapolis, IN, 1997.

Author Index

Clemensen, S., 309
Dukovska-Popovska, I., 97
Fujiwara, M., 229
Furukawa, H., 275
Gustaffson, M.G., 29
Hariyama, M., 265
Imaizumi, H., 37
Jorguseski, L., 157
Kameyama, M., 265
Koch, P., 211
Kramers, A., 29
Krigslund, R., 97
Larsen, P.G., 185
Lausdahl, K., 185
Lian, X., 287
Ligthart, L.P., 287
Lindgren, P., 309
Litjens, R., 157
Mägi, L.E., 29
Manev, B., 97
Mihovska, A., 131
Mizuike, T., 55
Morikaw, H., 37
Nakajima, N., 145
Nielsen, R., 131
Nikookar, H., 287

Ohmori, S., 1
Oostveen, J., 157
Pedersen, G.F., 97
Popovski, P., 97
Prasad, R., 1, 65, 131
Pruthi, P., 65
Ramareddy, K., 65
Ribeiro, A., 185
Rohde, J., 185
Rovsing, P.E., 185
Sasaki, M., 229
Saughaug, K., 309
Šimunić, D., 1
Skouby, K.E., 87
Sugiyama, K., 55
Takeoka, M., 229
Tanaka, H., 229
Taran, Y., 309
Toftegaard, T.S., 185
Umehira, M., 245
Wada, N., 275
Waseda, A., 229
Windekilde, I., 87
Wolff, S., 185
Zhang, H., 157

Subject Index

Adaptive OFDM, 287
battery powered devices, 211
business model, 309
capacity, 229
channel matrix, 229
clean energy, 55
climate change, 65
coherent communication, 229
coherent state, 229
communications infrastructure, 185
communications network, 245
convergence, 117
deep space optical communications, 229
dematerialization, 1
dynamic switching on/off, 157
ecology, 55
e-learning, 29
energy consumption, 131
energy efficiency, 75, 245
energy harvesting, 97, 211
energy management, 185
energy saving, 157
energy-efficient networking, 37
fiber fuse, 229
fluorescent lamp, 145
functional-unit allocation, 265
green business model, 75
green communications, 1, 65
green house gasses, 65
green ICT, 55, 97, 117
green initiatives, 75
green Internet, 37
green mobile, 75
green radio, 287
green world, 1

Helstom receiver, 229
high-level synthesis, 265
home automation, 185
homodyne detection, 229
ICT and the environment, 65
ICT, 75
IEEE 802.15.4, 87
innovation, 309
Intelligent Transport Systems (ITS), 29
Intelligent WiMAX (I-WiMAX), 287
interconnection complexity, 265
locationing, 287
low power, 131
low power networks, 87
memory allocation, 265
metric of energy efficiency, 245
multiprocessor systems, 211
mutual information, 229
network based business model, 309
network planning, 117
optical buffer, 275
optical communication, 229
optical fiber, 145
optical label processor, 275
optical packet switch, 275
optimization, 117
power saving, 55
power supply, 145
protocol interoperability, 185
quadrature amplitude and phase, 229
quantum collective decoding, 229
quantum computing, 229
quantum noise, 229
quantum probability amplitude, 229
resource optimization, 211

RFID, 97
scheduling, 211, 265
sensor network, 87, 145, 185
Shannon formula, 229
sleep mode, 245
smart antennas, 287
smart energy, 1
smart housing, 1
smart ICT, 1
smart radio, 287
smart transport, 1

sustainable city, 29
sustainable energy, 65
system on a chip (SoC), 87, 131
telecommunication operator, 55
telemedicine, 29
W/bit, 245
wide-colored optical packet, 275
wireless access, 157
wireless sensor networks, 97
zero-crossing demodulation (ZXD), 87
ZigBee, 87

About the Editors

Ramjee Prasad

Professor Dr. Ramjee Prasad, Fellow of IEEE (USA), The IET (UK) and IETE (India), has obtained B.Sc. Engineering in Electronics and Communication from the Bihar Institute of Technology, Sindri, India in 1968 followed by a M.Sc. Engineering from the Birla Institute of Technology (BIT), Ranchi, India in 1970 and a Ph.D. from BIT, India in 1979.

Ramjee Prasad is a world-wide established scientist, who has given fundamental contributions towards development of wireless communications. He achieved fundamental results towards the development of CDMA and OFDM, taking the leading role by being the first in the world to publish books in the subjects of CDMA (1996) and OFDM (1999). He is the recipient of many international academic, industrial and governmental awards and distinctions, huge number of books (more than 25), journals and conferences publications (together more than 750), a sizeable amount of graduated PhD students (over 60) and an even larger amount of graduated M.Sc. students (over 200). Several of his students are today worldwide telecommunication leaders themselves. Recently, under his initiative, international M.Sc. and PhD programmes have been started with the Sinhgad Technical Education Society in India, the Bandung Institute of Technology in Indonesia and with the Athens Information Technology (AIT) in Greece.

Ramjee Prasad has a long path of achievements until to date and a rich experience in the academic, managerial, research, and business spheres of the mobile and wireless communication area. Namely, he played an important role in the success that the Future Radio Wideband Multiple Access Systems (FRAMES) achieved. He was the leader of successful EU projects like the MAGNET and MAGNET Beyond, among others, as well as the driver of fruitful cooperation with companies in projects, like Samsung, Huawei, Nokia, Telenor, among others. He started as a Senior Research Fellow (1970–1972) and continued as an Assistant Professor (1972–1980) at the Birla Institute of Technology (BIT), Mesra, Ranchi, India. He was appointed as an Associate Professor in 1980–1983 and head of the Microwave Laboratory there. From 1983–1988 Ramjee Prasad worked at the University of Dar es Salaam (UDSM), Tanzania, where he became Full Professor of Telecommunications in the Department of Electrical Engineering in 1986.

From February 1988 till May 1999 Ramjee Prasad worked at the Delft University of Technology (DUT), The Netherlands at the Telecommunications and Traffic Control Systems Group. He was the founding head and program director of the Centre for Wireless and Personal Communications (CWPC) of the International Research Centre for Telecommunications-Transmission and Radar (IRCTR) at DUT, The Netherlands. Since June 1999, Ramjee Prasad has been holding the Professorial Chair of Wireless Information and Multimedia Communications at Aalborg University, Denmark (AAU). Here, he was also the Co-Director of the Center for PersonKommunikation until December 2002. He became the research director of the department of Communication Technology in 2003. In January 2004, he became the Founding Director of the Center for TeleInfrastruktur (CTIF), established as large multi-area research center at the premises of Aalborg University.

CTIF at Aalborg University was inaugurated on January 29, 2004. Under Ramjee Prasad's successful leadership and due to his extraordinary vision, CTIF turned into CTIF-Global by opening 4 divisions, namely: CTIF-Italy (inaugurated in 2006 in Rome), CTIF-India (inaugurated in 2007 in Kolkata), CTIF-Copenhagen and CTIF-Japan (inaugurated in 2008).

Ramjee Prasad is the founding chairman of Global ICT Standardization Forum for India (GISFI).

Shingo Ohmori
Professor Dr. Shingo OHMORI is a President and professor of CTIF-Japan, Aalborg University, Denmark. He received the B.E., M.E., and Ph.D. degrees in electrical engineering from the University of Tohoku, Japan, in 1973, 1975, and 1978, respectively. In 1978, he joined National Institute of Information and Communications Technology (NICT), and he resigned a Vice President of NICT in March 2009.

He has been engaged in research on mobile satellite communications, especially on antenna and propagation. During 1983–1984, he was a visiting research associate at the ElectroScience Laboratory, the Ohio State University, Columbus, Ohio.

He is the author of *Mobile Satellite Communications* (Artech House, 1998) and a co-author of *Mobile Antenna Systems Handbook* (Artech House,

1994). He was awarded the Excellent Research Prize from the Minister of Science and Technology Agency of Japan in 1985, and the Excellent Research Achievements Prize of the IEICE in 1993. Dr. Ohmori is an IEEE Fellow and an IEICE Fellow (Institute of Electronics, Information and Communication Engineers, Japan).

Dina Šimunić

Dr. Dina Šimunić is a full professor at University of Zagreb, Faculty of Electrical Engineering and Computing in Zagreb, Croatia. She graduated in 1995 from University of Technology in Graz, Austria. In 1997 she was a visiting professor in "Wandel & Goltermann Research Laboratory" in Germany, as well as in "Motorola Inc.", Florida Corporate Electromagnetics Laboratory, USA, where she worked on measurement techniques, later on applied in IEEE Standard. In 2003 she was a collaborator of USA FDA on scientific project of medical interference. Dr. Simunic is a IEEE Senior Member, and acts as a reviewer of *IEEE Transactions on Microwave Theory and Techniques*, *IEEE Transactions on Biomedical Engineering and Bioelectromagnetics*, the journal *JOSE* and as a reviewer of many papers on various scientific conferences (e.g., IEEE on Electromagnetic Compatibility). She was a reviewer of Belgian and Dutch Government scientific projects, as well as of the Fifth Framework EU programs. She was acting as a main organizer of the data base in World Health Organization, for the service of International EMF Project from 2000 to 2009. From 1997 to 2000 she acted as a vice-chair of COST 244: "Biomedical Effects of Electromagnetic Fields". From 2001 to 2004 she served as vice chair of Croatian Council of Telecommunications. In 2006 she is elected vice-chair of COST Domain Committee on Information and Communication Technologies (ICT). She is also a member of COST Transdomain Committee, which evaluates transdomain action proposals. She is organizer of many workshops, symposia and round tables (e.g., WHO Round Table on Standards Harmonization in Zagreb, Croatia, 1998), as well as of special sessions (e.g., special session on telemedicine and intelligent transport systems during Wireless Vitae, Aalborg, Denmark in 2009). She has held numerous invited lectures, among others at ETH Zuerich, Switzerland in 1996 and US Air Force, Brooks, USA in 1997). She is author or co-author of approximately 100 publications in various journals and books, as well as her student text for wireless communications, entitled: *Microwave Communications Basics*. She is a co-author of the book *Green Communication*, to be published in February 2010. In the previous work, Dr. Simunic won three times the first prize for the best young scientists work, and in the recent past, acted as a mentor of approximately 100 students in the field of wireless communications and electromagnetics, among whom several got the first prize on the University of Zagreb for their best student work. She is currently teaching "Theory of Wireless Communications Systems" and "Biomedical Effects of Electromagnetic Fields". She is also a lecturer in study on ICT security. Her research work (on the national level she had several scientific projects) comprises electromagnetic fields dosimetry and wireless communications.

About the Authors

Suberia Clemmensen
Suberia Clemmensen, Research Fellow, MSc Econ & Business Adm, Innovation, & Entrepreneurship. Her primary research interest is focussed on eco- or green-innovation for businesses, i.e., in transforming businesses to become sustainable by balancing their economic, environmental and social objectives. This research area builds on Suberia's work and academic experience in the field of sustainable community and regional development, and complements her demonstrated work experience with organisational change management.

Iskra Dukovska-Popovska
Iskra Dukovska-Popovska is an Assistant Professor at the Centre for Logistics, Aalborg University. She holds a B.Sc. in Mechanical Engineering and a M.Sc. in Industrial Engineering from the Faculty of Mechanical Engineering, University 'Sts. Cyril and Methodius' in Skopje, Macedonia, and a Ph.D. in Manufacturing Strategic Management from Aalborg University, Denmark. She has published articles in the field of Manufacturing strategy management, Supply chain strategy management, Interaction of strategic management, contingencies and performance, and Human aspects in strategy formation. She is a member of International Manufacturing Strategy Survey project. Besides the above mentioned areas, her research focus is extending towards environmental aspects of logistics and supply chain management, and the use of RFID in such context.

Mikio Fujiwara
Mikio Fujiwara received the B.S., M.S. degrees in electrical engineering from Nagoya University, Aichi Japan, in 1990, 1992, respectively. In 2002, he received Ph.D. degree in physics from Nagoya University. In 1992, He joined Communications Research Laboratory, Ministry of Posts and Telecommunications, and was engaged in the development of Ge: Ga

far-infrared photoconductors.

Since 2000, he has been in the quantum information technology group. His current interests include GaAs JFETs and InGaAs pin photodiodes for development of ultra-sensitive photo-detectors in the telecom-bands. Dr. Fujiwara is a member of the Japanese Society of Physics, and the Institute of Electronics, Information and Communication Engineers of Japan.

Hideaki Furukawa

Hideaki Furukawa received the B.E., M.E. and Dr. Eng. degrees in Material and Life Science from Osaka University, Osaka Japan, in 2000, 2002 and 2005, respectively.

Since 2005, he has been with the National Institute of Information and Communications Technology (NICT), Tokyo, Japan. His research interests include photonic information technology and photonic networks.

Dr. Furukawa is a member of the Institute of Electrical and Electronics Engineers (IEEE), the Institute of Electronics, Information and Communication Engineers (IEICE), and the Japan Society of Applied Physics (JSAP).

Matilda Gennvi Gustafsson

Matilda Gennvi Gustafsson worked with environmental and production management in the food industry during the early 1990s and joined Ericsson in 1997. Focusing over the years on management, business and change management, she is now responsible Sustainability Director within Ericsson.

Masanori Hariyama

Masanori Hariyama received the B.E. degree in electronic engineering, M.S. degree in Information Sciences, and Ph.D. in Information Sciences from Tohoku University, Sendai, Japan, in 1992, 1994, and 1997, respectively. He is currently an associate professor in Graduate School of Information Sciences, Tohoku University. His research interests include VLSI computing for real-world application such as robots, high-level design methodology for VLSIs and reconfigurable computing.

Hideaki Imaizumi

Dr. Imaizumi received his B.A. degree in Policy Management in 1995, M.A. and Ph.D. degrees in Media and Governance in 2001 and 2005 all from Keio University, Kanagawa, Japan. From 2002 to 2003, he stayed in University of Southern California, Information Science Institute (USC/ISI) as an exchange visitor. Since obtaining his Ph.D., he has begun to work as a research associate in the University of Tokyo, Japan. Since 2008, he has been an assistant professor in the Research Center for Advanced Science and Technology at the University of Tokyo. His current research interests are in the area of optical communication technologies, photonic network architecture, and green communication technology. He received the IEICE Young Researcher's Award in 2009. He is a member of IEEE, ACM, and IEICE.

Ljupčo Jorgušeski

Ljupčo Jorgušeski is a senior consultant/innovator in the area of wireless access networks. His current working responsibilities are: optimization and performance estimation of wireless access systems such as HSPA+, WiMAX and LTE, strategic technology assessment and knowledge development of future wireless access systems, and monitoring and active participation in 3GPP standardisation and NGMN. He holds a Dipl. Ing. degree (in 1996) from the Faculty of Electrical Engineering at the university "Sts. Cyril i Metodij", Skopje, Republic of Macedonia and a Ph.D. degree (in 2008) in wireless telecommunication from Aaalbog University, Aalborg, Denmark. From 1997 to 1999 he worked as applied scientist at TU Delft, Delft, the Netherlands on the ETSI project FRAMES that defined the European proposal for the UMTS standardization. From 1999 to 2003 he was with KPN Research, Leidschendam, the Netherlands where he was working on the radio optimization and performance estimation of the GPRS, UMTS and WLAN networks. He has published over 15 conference papers, one journal paper, two book chapters, and co-authored one patent (including six pending patent applications).

Peter Jung

Peter Jung received the diploma (M.Sc. equiv.) in physics from University of Kaiserslautern, Germany, in 1990, and the Dr.-Ing. (Ph.D. E.E. equiv.) and Dr.-Ing. habil. (D.Sc. E.E. equiv.), both in electrical engineering with focus on microelectronics and communications technology, from University of Kaiserslautern in 1993 and 1996, respectively. In 1996, he became private educator (equivalent to reader) at University of Kaiserslautern and in 1998 also at Technical University of Dresden, Germany. He left the University of Kaiserslautern in March 1998 and from March 1998 till May 2000 he was with Siemens AG, Bereich Halbleiter, now Infineon Technologies, as Director of Cellular Innovation and later Senior Director of Concept Engineering Wireless Baseband. In June 2000, he became Chaired

Professor for Communication Technologies (Kommunikationstechnik) at the Gerhard-Mercator-University Duisburg. His areas of interest include wireless communication technology, software defined radio, and system-on-a-chip integration of communication systems.

Michitaka Kameyama

Michitaka Kameyama received the B.E., M.E. and D.E. degrees in Electronic Engineering from Tohoku University, Sendai, Japan, in 1973, 1975, and 1978, respectively. He is currently a Professor in the Graduate School of Information Sciences, Tohoku University. His general research interests are intelligent integrated systems for real-world applications and robotics, advanced VLSI architecture, and new-concept VLSI including multiple-valued VLSI computing. Dr.Kameyama received the Outstanding Paper Awards at the 1984, 1985, 1987 and 1989 IEEE International Symposiums on Multiple-Valued Logic, the Technically Excellent Award from the Society of Instrument and Control Engineers of Japan in 1986, the Outstanding Transactions Paper Award from the IEICE in 1989, the Technically Excellent Award from the Robotics Society of Japan in 1990, and the Special Award at the 9th LSI Design of the Year in 2002. Dr. Kameyama is an IEEE Fellow.

Peter Koch

Peter Koch earned his MSc and PhD in Electronics Engineering in 1989 and 1996, respectively, both degrees from Aalborg University. Since 1997 he has been an Associate Professor in Digital Signal Processing and ASIC Design, and since mid 2006 he has been heading the Center for Software Defined Radio, CSDR, at Aalborg University, a regional technology center with the prime objective of conducting R&D cooperation with industry in order to improve the general skills and competences in the associated companies within the domain of software defined radio.

Peter Koch's research interests include methodologies for energy minimization in programmable single- and multiprocessor environments, HW/SW co-design tools, and techniques for matching the interaction between digital signal processing algorithms and real-time target architectures. He is the author or co-author of more than 30 scientific papers, book chapters, and technical reports.

He is currently supervising three PhD students who are working on (a) methodologies for dynamically reconfiguration of FPGA platforms, (b) design methodologies for resource optimal real-time architectures for software defined radios based on multi-rate digital signal processing, and (c) battery-aware multi-core task scheduling, respectively. Previously, he has supervised three PhD students to completion, as well as one Postdoc student.

Together with Professors Ramjee Prasad and Ole Bruun Madsen, both Aalborg University, in 2004 Peter Koch defined the overall research and administration strategy for the Center for TeleInFrastruktur, CTIF, a research and innovation entity at Aalborg University engaged in a variety of wired- and wireless communication research. CTIF was inaugurated January 2004, and had a total budget of 120 mio DKR for the first five years. Until June 2006, Peter Koch acted as co-director for CTIF, and in 2005 he represented CTIF in the VTU established steering committee for the "Teknologisk Fremsyn" for Mobile and Wireless Communication. Currently he is the coordinator of CTIF's thematic area "Embedded and Resource Optimal Communication".

In cooperation with Professor Kim G. Larsen and three other colleagues from Aalborg University, Peter Koch was one of the promoters of "Center for Indlejrede Software Systemer", CISS, a national competence center under the "Jysk-Fynske IT-satsning". CISS had a total budget of 62.5 mio. DKR during the period from 2002 to 2006. Until mid 2005, Peter Koch was a member of the CISS management board.

With Professor Kim G. Larsen, AAU, and Project Director Esben Wolf, Teknologisk Institut, in 2004 Peter Koch initiated and was granted the High Technology Network "Mobile Systems" funded by the Danish Ministry for Science, Technology and Innovation. The network had a total budget of 2×10 mio. DKR, and was actively running during 2005–2008.

Peter Koch is one of the founders of and main contributors to the MSc program "Applied Signal Processing and Implementation" (ASPI) which has run continuously since 1993 at Aalborg University. At this program, Peter Koch has supervised more than 20 MSc projects, and he was awarded "Best Teacher of the Year" at the Faculty of Nature and Science, Aalborg University, in 1994 as well as in 1998, in particular for his strong engagement in the ASPI program.

Anna Kramers
Anna Kramers has worked with developing multimedia business opportunities since the beginning of 2000. She has also studied industrial ecology at the Royal Institute of Technology in Stockholm. She is currently a Research Advisor in the filed of Sustainable Cities at KTH Center of Sustainable Communications.

Rasmus Krigslund

Rasmus Krigslund received the B.Sc. degree in Electrical Engineering and Information Technology and the M.Sc. (Cum Laude) in Wireless Communication Systems from Aalborg University, Aalborg, Denmark, in 2007 and 2009, respectively. From 2006 to 2009 he has been employed as a student helper at Aalborg University where he has participated in various research and administrative projects. Today he is pursuing a Ph.D. degree in micro localization in Radio Frequency Identification (RFID) systems and is regularly used as a reviewer for IEEE Communication Letters. His research interests are in the area of wireless communication and networking in general.

Peter Gorm Larsen

Peter Gorm Larsen studied computer science at the Technical University of Denmark (DTU). He received his MSc in 1988 and PhD in 1995, with focus on semantics, computer languages and tool support. He has worked in industry for most of his career. For 13 years he worked with IFAD and was the main architect of VDMTools and he was responsible for support of VDMTools world-wide. For more than three years he worked for Systematic Software Engineering mainly doing business development for large defence projects. He is now a full professor at Aarhus School of Engineering, Denmark.

Kenneth Lausdahl

Kenneth Lausdahl has a M.Sc. in ICT engineering from Aarhus School of Engineering, Denmark with focus on distributed systems and tool support. He has since worked at Aarhus School of Engineering, Denmark in the domotics field, low powered wireless networks and software development. He has published two articles at international conferences.

Rami Lee

Rami Lee received her B.Sc. diploma from Yonsei University, Korea, in 1999 and the M.Sc. diploma from POSTECH, Korea, in 2001, both in electrical engineering with wireless communications technology. From 2001 until now she has been with SK Telecom, Korea as project manager such as wireless commucation engineering and ZigBee chip development.

Xiaohua Lian

Xiaohua Lian received the B.S. and M.S. degrees in electrical and electronic engineering from Nanjing University of Aeronautics and Astronautics

(NUAA), Nanjing, China, in 2002 and 2005 respectively. She is now working towards her Ph.D. at the International Research Centre for Telecommunications and Radar (IRCTR) of the Department of Electrical Engineering, Mathematics and Computer Science of TUDelft. Her areas of interest include signal processing for wireless communications, smart antenna and cognitive radio, etc.

Leo P. Ligthart

Leo P. Ligthart was born in Rotterdam, the Netherlands, on September 15, 1946. He received an Engineer's degree (cum laude) and a Doctor of Technology degree from Delft University of Technology in 1969 and 1985, respectively. He is fellow of IET and IEEE. He received Doctorates (honoris causa) at Moscow State Technical University of Civil Aviation in 1999 and Tomsk State University of Control Systems and Radioelectronics in 2001. He is academician of the Russian Academy of Transport. Since 1992, he has held the chair of Microwave Transmission, Radar and Remote Sensing in the Department of Electrical Engineering, Mathematics and Computer Science, Delft University of Technology. In 1994, he founded the International Research Center for Telecommunications and Radar (IRCTR) and is the director of IRCTR. Prof. Ligthart's principal areas of specialization include antennas and propagation, radar and remote sensing. However, he has also been active in satellite, mobile and radio communications. He has published over 500 papers and 2 books.

Peter Lindgren

Peter Lindgren, Associated Professor, PhD is Associated Professor of Innovation and New Business Development at the Center for Industrial Production at Aalborg University. He is manager of International Center for innovation and he holds a B.Sc. in Business Administration and in M.Sc. in Foreign Trade and Ph.D. both in Network based High Speed Innovation. He has (co-)authored numerous articles and several books on subjects such as Product Development in Network, Electronic Product development, New Global Business Development, Innovation Management and Leadership and High Speed Innovation. His current research interest is in New Global Business models, i.e. the typology and generic types of business models and how to innovate these.

Remco Litjens

Remco Litjens is a senior scientist at TNO Information and Communication

Technology, The Netherlands, with a focus on planning and optimisation of mobile cellular networks. Remco has over 12 years experience in technical consultancy, applied scientific research and training, regarding the development, assessment, planning, optimisation, performance and capacity management of diverse mobile access technologies. He holds Master degrees in Econometrics (1994, Tilburg University, the Netherlands) and Electrical Engineering and Computer Science (1996, UC Berkeley, USA) and a Ph.D. degree in Applied Mathematics (2003, University of Twente, the Netherlands).

Ole Brun Madsen

Ole Brun Madsen is a Professor in Distributed Real-time Systems and has been the Head of CNP (Center for Network Planning) since 2004. He is Co-director for CTIF, Center for TeleInFrastruktur at Aalborg University and from 2004 to 2009 he was the head of NetSec, Networking and Security section. He received his M.Sc. in Mathematics & Computer Science from the University of Copenhagen and worked as a researcher at and later the head of the Computer Science Laboratory at The Royal Danish Academy of Fine Arts in Copenhagen (1962–1972). He was the head of the Development Department, RECAU, the Regional Computing Centre at Århus University (1972–1981) and the head of the Data Network Section and the Network Infrastructure Strategy section at Jutland Telephone (1981–1996). From 1996 to 1999 he was Manager for Infrastructure Network Technology and Strategy at TDC, Tele Denmark. He has been project leader for a number of national and international R&D projects and acted in high level advisory tasks within the European Commission on the R&D framework programs in DGXIII and with United Nations UNDP activities. Present research is focused on Infrastructure Architecture and Modeling Tools for Network Analysis and Design.

Linda Ekener Mägi

Linda Ekener Mägi holds a Master of Science in Computer Science from Chalmers Technical University and a Bachelor of Business Administration from Gothenburg University. She has been working in the telecoms area for more than 10 years in various product management, business development, marketing and sales positions. She is now Director, Sustainability Communication within Ericsson.

Boris Manev

Boris Manev received his degree in International Economics and International Communications from the American University of Paris in 2008, in Paris, France. He was a scholarship student of the Annenberg Foundation for studies of global communications. Currently he is pursuing his MSc degree in Environment and Resource Management, with specialization in Energy Management, at the Vrije Universiteit Amsterdam, in Amsterdam, Netherlands.

From 2009 to 2010, Boris worked in the EU energy and environment field in Brussels, Belgium, as an advisor for Bellona Europa aisbl on topics related to CO_2 Capture and Storage technologies and the EU energy policy with respect to the Arctic.

Boris is the author of the book *Waste of Food* and the founder of the American University of Paris journal in global communications *Convergence*. He also published several articles in academic journals and NGO websites. His research interests are in CO_2 neutral technologies.

Albena Mihovska

Albena Mihovska (B.Sc., M.Sc. (1999), PhD (2009, Aalborg University)) is currently an Associate Professor at the Center for TeleInFrastruktur (CTIF) at Aalborg University in the area of next generation communication systems, network and radio resource management. She has a strong research background in next generation system concept design and is currently involved with advanced system concepts for LTE-Advanced and IMT-Advanced. Albena Mihovska was the coordinator of the FP6 SSA project SIDEMIRROR and FP5 project PRODEMIS and she has more than 50 publications, including four books (Artech House, 2009).

Takeshi Mizuike

Takeshi Mizuike received the B.S. degree in Electrical Engineering from University of Tokyo in 1977. He received the M.S. degree in Operations Research from Stanford University in 1981. In 1999, he also received his Dr. Eng. Degree from University of Tokyo. In 1977, he joined R&D Laboratories of Kokusai Denshin Denwa (KDD) Co., Ltd. (currently KDDI Corporation), where he had been engaged in the research on wireless communication networks and research management on network planning techniques. From 1990 to 1992, he worked at Engineering Department of INTELSAT headquarters in Washington, DC, USA. Since 1999, he worked for management of KDDI R&D Laboratories as Vice President, Managing

Director and for technology planning at Technology Strategy Department of KDDI as General Manager. He is currently Vice President, Chief Executive Director of Technology Development Center of KDDI R&D Laboratories. Dr. Mizuike is a Fellow of the Institute of Electronics, Information and Communication Engineers (IEICE) of Japan and a Senior Member of the IEEE.

Hiroyuki Morikawa

Hiroyuki Morikawa received the B.E., M.E, and Dr. Eng. degrees in electrical engineering from the University of Tokyo, Tokyo, Japan, in 1987, 1989, and 1992, respectively. Since 1992, he had been in the University of Tokyo and is currently a full professor of the Research Center for Advanced Science and Technology at the University of Tokyo. From 1997 to 1998, he stayed in Columbia University as a visiting research associate. From 2002 to 2006, he was a group leader of NICT Mobile Networking Group. His research interests are in the areas of computer networks, ubiquitous networks, mobile computing, wireless networks, photonic Internet, and network services. He served as a technical program committee chair of many IEEE/ACM conferences and workshops, Director of IEICE, Editor-in-Chief of IEICE Transactions of Communications, and sits on numerous telecommunications advisory committees and frequently serves as a consultant to government and companies. He received more than 20 awards including the IEICE best paper award in 2002 and 2004, IPSJ best paper award in 2006, Info-Communications Promotion Month Council President Prize in 2008, NTT DoCoMo Mobile Science Award in 2009. He is a fellow of IEICE, and a member of IEEE, ACM, ISOC, IPSJ, and ITE.

Nobuo Nakajima

Nobuo Nakajima received the B.S., M.S. and Ph.D degrees in electrical engineering from Tohoku University, Sendai, Japan, in 1970, 1972 and 1982, respectively.

In 1972 he joined the Electrical Communication Laboratory, NTT. From 1972 to 1979, he was engaged in the research on millimeter-wave circuits. From 1980 to 1985, he was working under the development of microwave and mobile radio antennas. After 1985, he was engaged in the system design of the digital cellular communication system.

In 1992, he moved to NTT DoCoMo and in 1998, he became a senior vice president. During in NTT DoCoMo, he was engaged in the development of future mobile communication systems such as IMT-2000 and 4th generation system.

In 2000, he moved to University of Electro-Communications and he is a professor of the department of human communications and Advanced Wireless Communication Research Center.

Alberto Nannarelli

Alberto Nannarelli received his BS degree in electrical engineering from the University of Rome La Sapienza, Italy, in 1988 and the MS and PhD degrees in electrical and computer engineering from the University of California, Irvine, in 1995 and 1999, respectively. He is an associate professor at the Technical University of Denmark. He worked for SGS-Thomson Microelectronics and for Ericsson Telecom as a design engineer and for Rockwell Semiconductor Systems as a summer intern. From 1999 to 2003, he was with the Department of Electrical Engineering, University of Rome Tor Vergata, Italy, as a postdoctoral researcher. His research interests include computer arithmetic, computer architecture, and VLSI design.

Rasmus Hjorth Nielsen

Rasmus Hjorth Nielsen (B.Sc., M.Sc. (2005), Ph.D. (2009, Aalborg University)) is currently working as a PostDoc at the Center for TeleInFrastruktur (CTIF) at Aalborg University. He is working within the field of next generation networks where his focus is security and performance optimization. He has a strong background in operational research and optimization in general and has applied this as a consultant within planning of large-scale networks. He is a partner and member of the board in a company providing such expertise.

Homayoun Nikookar

Homayoun Nikookar received his Ph.D. in Electrical Engineering from Delft University of Technology (TUDelft), The Netherlands, in 1995. He is an Associate Professor at the International Research Centre for Telecommunications and Radar (IRCTR) of the Department of Electrical Engineering, Mathematics and Computer Science of TUDelft. He is also the leader of the Radio Advanced Technologies and Systems (RATS) program of IRCTR, leading a team of researchers carrying out cutting edge research in the field of radio transmission. He has conducted active research in many areas of wireless communications, including wireless channel modeling, UWB, MIMO, multicarrier transmission, Wavelet-based

OFDM and Cognitive Radio. He is the co-recipient of the 2000 IEEE Neal Shepherd Best Propagation Paper Award for publication in the March issue of Transactions on Vehicular Technology. He is also a recipient of several paper awards of International Conferences and Symposiums. In 2007 Dr Nikookar served as the Chair of the 14th IEEE Symposium on Communications and Vehicular Technology (SCVT) in Benelux and in 2008 was the Chairman of the European Wireless Technology Conference (EuWiT). His recent paper, Cognitive Radio Modulation Techniques, *IEEE Signal Processing Magazine*, November 2008, has been listed in the IEEE Xplore top 100 documents accessed in January and February 2009. Dr. Nikookar is a Senior Member of the IEEE and the coauthor of the book, *Introduction to Ultra Wideband for Wireless Communications*, Springer, 2009.

Job Oostveen

Job Oostveen is a senior researcher and technical consultant at TNO Information and Communication Technologies. His consultancy work touches both technology innovation in cellular networks and more generic aspects of radio communications. His research is focused on performance assessment, radio network planning (incl. propagation) and optimization of MIMO-OFDM-based mobile networks. Among others, he is involved in radio network standardization in 3GPP. Before joining TNO, he was affiliated with Philips Research, where his research was focused on signal processing for multimedia (mainly video watermarking and fingerprinting) and wireless communications (MIMO, OFDM, 802.11n). Job Oostveen holds a Ph.D. and M.Sc. in applied mathematics of the universities of Groningen and Twente, respectively.

Gert Frølund Pedersen

Gert Frølund Pedersen received the B.Sc. E.E. degree, with honour, in electrical engineering from College of Technology in Dublin, Ireland, and the M.Sc. E.E. degree and Ph.D. from Aalborg University in 1993 and 2003. He has been employed by Aalborg University since 1993 where he is now Professor for the Antenna, Propagation and Networking group. His research has focused on radio communication for mobile terminals including Antennas, Diversity systems, Propagation and Biological effects and he has published more than 60 peer reviewed papers and holds 17 patents. He has also worked as consultant for developments of antennas for mobile terminals including the first internal antenna for mobile phones in 1994 with very low SAR, first internal triple-band antenna in 1998 with low SAR and high efficiency, and lately various multi antenna systems rated as the most efficient on the market.

He has been one of the pioneers in establishing over-the-air measurement systems. The measurement technique is now well established for mobile terminals with single antennas and he is now chairing the COST2100 SWG2.2 group with liaison to 3GPP for over-the-air test of MIMO terminals.

Petar Popovski

Petar Popovski received the Dipl.-Ing. in electrical engineering and M.Sc. in communication engineering from the Faculty of Electrical Engineering, Sts. Cyril and Methodius University, Skopje, Macedonia, in 1997 and 2000, respectively and a Ph.D. degree from Aalborg University, Denmark, in 2004. He worked as Assistant Professor at Aalborg University from 2004 to 2009. From 2008 to 2009 he held part-time position as a wireless architect at Oticon A/S. Since 2009 he is an Associate Professor at Aalborg University. He has more than 90 publications in journals, conference proceedings and books and has more than 20 patents and patent applications. In January 2009 he received the Young Elite Researcher award from the Danish Ministry of Science and Technology. He has received several best paper awards, among which the ones at IEEE Globecom 2008 and 2009. He has served as a technical program committee member in various conferences and workshops, including IEEE SECON, IEEE ICCCN, IEEE PIMRC, IEEE Globecom, etc. Dr. Popovski has been a Guest Editor for special issues in *EURASIP Journal on Applied Signal Processing* and the *Journal of Communication Networks*. He serves on the editorial board of *IEEE Transactions on Wireless Communications, IEEE Communications Letters, Ad Hoc and Sensor Wireless Networks* journal, and *International Journal of Communications, Network and System Sciences* (IJCNS). His research interests are in the broad area of wireless communication and networking, information theory and protocol design.

Parag Pruthi

Dr. Parag Pruthi is the founder, Chairman and CEO of NIKSUN, Inc. He brings over twenty three years of expertise in network security, surveillance, data warehousing/mining and systems performance evaluation. Based upon his fundamental research on the use of chaos theory to model high variability phenomenon in networking, Dr. Pruthi along with notable colleagues in the industry developed unique methods of analyzing network traffic enabling a scalable and integrated approach to network security and performance. At

NIKSUN, Dr. Pruthi drives the vision and strategic thinking for the company. His goals are to create a unified and holistic solution for end-to-end network security while guaranteeing end-to-end service delivery requirements.

Kishore Ramareddy

Kishore Ramareddy is Senior Software Engineer at NIKSUN Inc. He has thirteen years of expertise in areas of network security, network management, data mining and data warehousing. His research interests areas include network monitoring, network security and artificial intelligence.

Sebastian Rickers

Sebastian Rickers studied Electrical and Information Engineering at the University of Duisburg-Essen with a specialization on communication technologies. After a three-month internship at Siemens in Minnetonka, USA, he returned to Germany to complete his studies in 2009 with the diploma degree (M.Sc. EE equiv.). Since May 2009 he has been with the Chair of Communication Technologies at the University of Duisburg-Essen. His focus is on ZigBee technologies.

Augusto Ribeiro Augusto Ribeiro has a master degree in computer science completed Minho University in Portugal and focus on formal methods. He has since worked for one year in Aarhus School of Engineering, Denmark in the domotics field, low power wireless networks and software development.

John Rohde

John Rohde received his M.Sc.EE (1992) in microwave electronics from the Technical University of Denmark (DTU) in Lyngby, Denmark. Fields of expertise and main research activities are within the area of RF front-end design including LNA, PA and antennas for wireless communications. John has 5+ years of working experience within these areas from previous employments as research assistant at the Electromagnetics Institute, DTU, and as RF development Engineer at Freescale Semiconductor and Develco A/S. Currently he holds a position as Associate Professor at Aarhus School of Engineering, Denmark.

Kristin Falck Saughaug

Kristin Falck Saughaug, Research Fellow, MA Theology, is a research fellow at the Center for Industrial Production at Aalborg University. Her current research interest lies in the area of the creating human being.

Focus is the potential of innovation defined via the connection of different knowledge and business domains. Other areas of research interests include innovation leadership, intellectual capital management, human resource and organizational theory, artteori, philosophic theology and ethics.

Woojin Shim
Woojin Shim received his Bachelor's degree from Vanderbilt University, USA, in 2001, and M.Sc. in Optical Communication from Yonsei University of Seoul, Korea in 2005. He has been employed for SK Telecom of Korea from 2005 until now, working for the Institute of Future Techonology in the Ambient Techonology Development team. He is involved and working as project manager in WPAN projects such as ZigBee Chip development and mesh network technology development. He has also been engaged in ISO as a Korean Delegate, from 2007 until now, to standardize novel mesh network technology.

Knud Erik Skouby
Knud Erik Skouby is professor and founding director of center for Communication, Media and Information technologies (CMI), Aalborg University, Copenhagen – a center providing a focal point for multi-disciplinary research and training in applications of CMI – as part of the international research network CTIF. He has a career within consultancy and as a university teacher since 1972. The working areas have for the last 20 years been within the telecom area focusing on mobile/wireless: Techno-economic analyses; Development of mobile/wireless applications and services: Regulation of telecommunications Chair of WG1 in WWRF – Wireless World Research Forum – a forum for pre-standardization activities. He has participated as a project manager and partner in a number of international (including a long term activity in Africa), European and Danish research projects. He has served on a number of public committees within the areas of telecom, IT and broadcasting; as a member of boards of professional societies; as a member of organizing boards, evaluation committees and as invited speaker on international conferences; published a number of Danish and international articles, books and conference proceedings in the areas of convergence, mobile/wireless development, telecommunications regulation, technology assessment (information technology and telecommunications), demand forecasting and political economy.

Christoph Spiegel

Christoph Spiegel has been with the Department of Communication Technologies at the University of Duisburg-Essen since 2006. He received the diploma degree (M.Sc. equiv.) in Electrical and Information Engineering with focus on communication technologies in 2005 from the University of Duisburg-Essen. He is involved in several project groups such as 3G evolution, software defined radio techniques, digital video broadcasting and ZigBee technologies.

Martin Hjort Stender

Martin Hjort Stender holds a M.Sc.E.E. (1995) in wireless communication from Aalborg University. He started his professional carrier in the telecommunication industry developing protocol stack software for DECT, GSM, UMTS and VoIP. The last three years he has worked as an independent engineering consultant for companies like Texas Instruments and RTX Telecom. Today Martin is working as an Associate Professor at the Aarhus School of Engineering, Denmark.

Keizo Sugiyama

Keizo Sugiyama received his B.S., M.S., and Ph.D. degrees in Information Science from Kyoto University, Kyoto, Japan, in 1985, 1987 and 2004, respectively. He joined Kokusai Denshin Denwa Co., Ltd. (now KDDI Corp.), in 1987. Since 1987, he has been engaged in research on OSI protocols, Electronic Data Interchange (EDI), network management, Intelligent Transport Systems (ITS) and Wireless LAN. He is currently a senior manager in Technology Planning and Evaluation Group of Technology Development Center of KDDI R&D Laboratories. He received the Young Researcher's Award from IEICE in 1995.

Masahiro Takeoka

Masahiro Takeoka received a Ph.D. degree in electrical engineering from Keio University in 2001. Since 2001, he has been in National Institute of Information and Communications Technology (NICT) where he is currently a senior researcher and working on theory and proof-of-principle experiments on quantum detection, quantum information, and quantum optics. He is a member of the Physical Society of Japan.

Hidema Tanaka

Hidema Tanaka received the B.E., M.E., and Ph.D. degrees all in Electrical

Engineering, from Science University of Tokyo, in 1995, 1997 and 2000, respectively. In 2000, he joined the faculty of Science and Technology, Science University of Tokyo. In 2002, he joined Communication Research Laboratory. Currently, he is a senior researcher of Security Fundamentals Group in National Institute of Information and Communications Technology. His interests include information theory, code theory, information security and cryptanalysis. He is a member of the Institute of Electronics, Information and Communication Engineers of Japan.

Yariv Taran

Yariv Taran, PhD Student, M.Sc. Economics and Business Administration, is a research fellow at the Center for Industrial Production at Aalborg University. His current research interest lies in the area of business model innovation. Other areas of research interests include intellectual capital management, knowledge management, entrepreneurship and regional systems of innovation.

Thomas Skjødeberg Toftegaard

Thomas Skjødeberg Toftegaard (former Thomas Toftegaard Nielsen) holds a M.Sc.E.E. (1995) and a Ph.D. (1999) in wireless communications from Aalborg University, Denmark. He holds a position as Professor in Communication Technology at Aarhus School of Engineering, Denmark. Furthermore, he serves as Director of the Electrical Engineering and Information & Communication Technology at Aarhus School of Engineering, Denmark.

Masahiro Umehira

Masahiro Umehira received the B.E., M.E. and Ph.D. degrees from Kyoto University in Japan in 1978, 1980 and 2000, respectively. He joined NTT (Nippon Telegraph and Telephone Corporation) in 1980, where he was engaged in the research and development of modem and TDMA equipment for satellite communications, TDMA satellite communication systems, wireless ATM and broadband wireless access systems for mobile multimedia services and ubiquitous wireless systems. From 1987 to 1988, he was with the Communications Research Center, Department of Communications, Canada, as a visiting scientist. He has been active in the standardization related to wireless LAN and wireless PAN in Japan, Europe and USA. Since 2006, he has been a Professor of Ibaraki University, Hitachi-shi, Ibaraki Prefecture. His research interest includes broadband wireless access technologies, wireless networking and cognitive radio technologies for future fixed/nomadic/mobile wireless systems, future satellite communication systems and ubiquitous services using wireless technologies. He received Young Engineer award and Achievement award from IEICE in 1987 and 1999, respectively. He also received Education, Culture, Sports, Science and Technology Minister Award in 2001, TELECOM System Technology Award from the Telecommunications advancement Foundation in 2003. and the Outstanding Paper Award in Wireless VITAE 2009. He is a member of IEEE and a Fellow of IEICE, Japan.

Naoya Wada

Naoya Wada received the B.E., M.E., and Dr. Eng. degrees in electronics from Hokkaido University, Sapporo, Japan, in 1991, 1993, and 1996, respectively. In 1996, he joined the Communications Research Laboratory (CRL), Ministry of Posts and Telecommunications, Tokyo, Japan. He is currently a Senior Researcher of the National Institute of Information and Communications Technology (NICT), Tokyo, Japan. He has been project reader of Photonic Node Project and research manager of the Photonic Network Group from 2006. Since April 2009, he has been group reader of the Photonic Network Group.

His current research interests are in the area of photonic networks and optical communication technologies, such as optical packet switching (OPS) network, optical processing, and optical code-division multiple access (OCDMA) system. He has published more than 75 papers in refereed journals and more than 220 papers in refereed international conferences.

Dr. Wada received the 1999 Young Engineer Award from the Institute of Electronics, Information and Communication Engineers of Japan (IEICE), and the 2005 Young Researcher Award from the Ministry of Education, Culture, Sports, Science and Technology. He is a member of IEEE Comsoc, IEEE LEOS, IEICE, the Japan Society of Applied Physics (JSAP), and the Optical Society of Japan (OSJ).

Atsushi Waseda

Atsushi Waseda received the M.S., and Ph.D. degrees in information science from Japan Advanced Institute of Science and Technology, in 2002 and 2007, respectively.

Currently, he is an expert researcher of Security Fundamentals Group in National Institute of Information and Communications Technology. His interests include quantum security and cryptanalysis.

Iwona Windekilde

Iwona Windekilde is employed as a Postdoc researcher at Center for Communication, Media and Information Technologies, Copenhagen Institute of Technology/Aalborg University. She holds a Doctor of Science degree in Economics. Research areas of interest: Economy, Green ICT, Business Models, Personal Networks, Personal Services, Marketing, Management.

She completed her Ph.D. at Szczecin University in Poland in 2002. From the year 2002 to 2005 she worked as an Assistant Professor at Szczecin University in Poland at the Department of Economics and Organization of Telecommunication. During her work at the University she received a prize of the Headmaster of the Szczecin University for outstanding achievements on the field of didactics. In 2005 she received Marie Curie Intra-European Fellowships – project co-funded by the European Commission within the Sixth Framework Programme. The main objective of her project was research work on "The role of Convergence in Telecommunication in Development of Consumption of Information Services – Comparative Analysis of International Experiences". In 2006 she was employed at Center for Information and Communication Technologies, Technical University of Denmark as a Post Doc researcher in the area of business models for personal electronic networks.

Sune Wolff

Sune Wolff holds a M.Sc. (2009) in Distributed Real-time Systems from Aarhus School of Engineering, Denmark. From February 1st, 2009 he has been working at Aarhus School of Engineering as a technical developer on the research project "Minimum Configuration – Home Automation". His main research interests are on modelling and simulation of embedded systems, and he will initiate a Ph.D. project within this field beginning of 2010.

Jae Hwang Yu

Jae Hwang Yu received the B.Sc. degree in electronics engineering from Kyungpook National University, Taegu, Korea, in 1984 and the M.Sc. degree in electronics engineering from Yonsei University, Seoul, Korea, in 1986. In 2005 he received the Ph.D. from in Korea Advanced Institute of Science & Technology, Taejon, Korea. From 1988 to 1993, he was the senior researcher of electronic business division (RF & Microwave technology) of Kukje Corporation. Currently he is working for SK Telecom as team leader of the Ambient Technology Development Team, which is part of the Institute of Future Technology.

Haibin Zhang

Haibin Zhang is a scientist at TNO Information and Communication Technology in Delft, the Netherlands. He received his PhD degree in electrical engineering from Shanghai Jiao Tong University (SJTU), China, in 2002. He joined the Department of Electronic Engineering, SJTU, as a research assistant in 1996, where he became an associate professor in signal processing for wireless communications in 2002. During 2006–2007, he was working in Philips Research in Eindhoven, The Netherlands, as a Marie Curie Fellow awarded by the European Commission under the sixth framework program (FP6). His current research interests lie in the area of wireless communications, with the focus on green ICT, dynamic radio resource management, wideband channel modelling, radio planning and optimisation. He has published one book on OFDM (the only author), and is author or co-author of 25 journal papers and 11 conference papers. He is inventor of more than 30 granted and filed patents. Haibin is a senior member of IEEE, and serves as regular reviewer for international top journals and conferences.

RIVER PUBLISHERS SERIES IN COMMUNICATIONS

Volume 1
4G Mobile & Wireless Communications Technologies
Sofoklis Kyriazakos, Ioannis Soldatos, George Karetsos
September 2008
ISBN: 978-87-92329-02-8

Volume 2
Advances in Broadband Communication and Networks
Johnson I. Agbinya, Oya Sevimli, Sara All, Selvakennedy Selvadurai, Adel Al-Jumaily,
Yonghui Li, Sam Reisenfeld
October 2008
ISBN: 978-87-92329-00-4

Volume 3
Aerospace Technologies and Applications for Dual Use A New World of Defense and
Commercial in 21st Century Security
General Pietro Finocchio, Ramjee Prasad, Marina Ruggieri
November 2008
ISBN: 978-87-92329-04-2

Volume 4
Ultra Wideband Demystified Technologies, Applications, and System Design
Considerations
Sunil Jogi, Manoj Choudhary
January 2009
ISBN: 978-87-92329-14-1

Volume 5
Single- and Multi-Carrier MIMO Transmission for Broadband Wireless Systems
Ramjee Prasad, Muhammad Imadur Rahman, Suvra Sekhar Das, Nicola Marchetti
April 2009
ISBN: 978-87-92329-06-6

Volume 6
Principles of Communications: A First Course in Communications
Kwang-Cheng Chen
June 2009
ISBN: 978-87-92329-10-3

Volume 7
Link Adaptation for Relay-Based Cellular Networks
Başak Can
November 2009
ISBN: 978-87-92329-30-1

Volume 8
Planning and Optimisation of 3G and 4G Wireless Networks
J.I. Agbinya
January 2010
ISBN: 978-87-92329-24-0